William Buckler

The Larvae of the British Butterflies and Moths

Vol. VI (The Third and Concluding Portion of the Noctuae)

William Buckler

The Larvae of the British Butterflies and Moths
Vol. VI (The Third and Concluding Portion of the Noctuae)

ISBN/EAN: 9783743426184

Manufactured in Europe, USA, Canada, Australia, Japa

Cover: Foto ©berggeist007 / pixelio.de

Manufactured and distributed by brebook publishing software (www.brebook.com)

William Buckler

The Larvae of the British Butterflies and Moths

THE LARVÆ

OF THE

BRITISH BUTTERFLIES

AND

MOTHS.

BY

(THE LATE)

WILLIAM BUCKLER.

EDITED BY

GEO. T. PORRITT, F.L.S.

VOL. VI.

(THE THIRD AND CONCLUDING PORTION OF THE NOCTUÆ.)

LONDON:

PRINTED FOR THE RAY SOCIETY.

MDCCCXCV.

PRINTED BY ADLARD AND SON.
BARTHOLOMEW CLOSE, E.C., AND 20, HANOVER SQUARE, W.

PREFACE.

It was with the greatest reluctance that I undertook the editing of this work in succession to the late Mr. H. T. Stainton. My time was so completely occupied with other matters that at first I felt it would be impossible for me to accede to the request of the Ray Society; and it was only when it seemed likely that the further publication of the work might otherwise be indefinitely delayed, together with a promise from my friend Mr. W. Denison Roebuck, of Leeds, that he would assist me by copying out the manuscript for the letterpress, etc., that I consented.

The present volume completes Mr. Buckler's descriptions and figures of the Noctuæ. There were drawings of nearly all the species, but in the case of many no accompanying descriptions. Fortunately these omissions were almost entirely of species well known, and described in all histories of the British Lepidoptera, and which Mr. Buckler had evidently not considered it worth while to re-describe. For the same reason (as in the last volume) it has not been thought necessary to introduce fresh descriptions here.

Considerable delay was again occasioned through

the artist's inability (through illness, etc.) to proceed
sufficiently fast with the colouring of the plates, and
eventually the difficulty had to be overcome by the
employment of Mr. F. W. Frohawk, whose beautiful
work during the last few years in the delineation of
larvæ is so well known.

GEO. T. PORRITT.

CROSLAND HALL, HUDDERSFIELD;
February, 1895.

CONTENTS OF VOL. VI.

CLASSIFIED LIST OF THE SPECIES

IN THIS VOLUME.

HETEROCERA.

THE LARVÆ

OF THE

BRITISH MOTHS.

EREMOBIA OCHROLEUCA.

Plate LXXXVII, fig. 1.

ON the 22nd June, 1870, in striking at a specimen of *Lycæna alsus* in an old chalk-pit, I took in my net by chance a very delicate-looking, active, *Noctua* larva, which was quite a stranger to me. Believing I had obtained it from *Anthyllis vulneraria*, I put some of this plant with the larva into a box; but, on looking at it late in the evening, I saw it had not eaten any of the vetch, and seemed eager to escape.

As I could remember nothing but grass besides the *Anthyllis* growing in the spot where it was taken, I went out in the twilight and gathered a little of the first species of grass that came to hand, without noticing what it was.

Next morning I was very pleased to see that the larva had partaken freely of the grass; and having by me at this time, potted, a growing tuft of *Nardus stricta* —a species I had noted on the dry grassy slopes where I had been the day before, I too hastily assumed this to be the proper food of my captive, and placed it thereon, securing it with a glass cylinder. As my attention at this time was fully taken up by

EREMOBIA OCHROLEUCA.

many other larvæ, I forgot to look at my unknown for some days; and when I saw it again it was not the least grown, nor did it look well. This made me resort to other grasses, but without effect; and I had the mortification of seeing it, day by day, become smaller and feebler, till on the 2nd of July it died. But before the breath was quite out of its shrunken body my regrets were banished, thanks to Mr. W. H. Harwood, of Colchester, who sent me on July 1st a larva precisely similar in form and colour, but much larger in size; and, what was still better, feeding away hopefully on its proper food—the seeds in a panicle of cock's-foot grass (*Dactylis glomerata*).

In order to make quite sure of this being its proper food, I gathered fresh panicles of this grass, as well as of two or three other kinds, and put them in with the larva; but I saw that it roamed over the other kinds till it found the seeds of the cock's-foot grass, and then attacked them ravenously, thus perfectly satisfying every doubt. On the 3rd of July it retired to earth; and on the 30th the perfect insect came forth.

This full-grown larva varied in no respect from that which I had myself taken, save in size, for it was twice as large. It was one inch and a half in length, cylindrical, of moderate and uniform stoutness throughout, including the head, the lobes of which were rounded and full; the legs and pro-legs all well developed.

Its ground-colour was a bright but very pale opaque whitish-green; the very broad dorsal stripe whitish, the subdorsal stripe similar, but a trifle less in breadth. Between this and the spiracles the ground-colour became a little deeper; was bordered along the spiracles by a narrow stripe of full deep green; the subspiracular inflated stripe whitish; the belly and legs of the ground-colour, a trifle darker than the back. The head was also of the pale ground-colour, with a blackish streak across the mouth, and was more polished than the surface of the body, though that was rather glossy; the folds of the segmental divisions appeared white; the spiracles were black, as well as all the tubercular dots, which were plainly visible in their usual situations, those on the back smaller than the others, and every one of them furnished with a fine whitish hair; the anterior legs also spotted with black. (W. B., November 25th, 1870; E.M.M., June, 1871, VIII, 21.)

DIANTHÆCIA ALBIMACULA.

On the 19th July, 1873, I had the pleasure of receiving from Mr. Stainton five larvæ in different stages of growth, which had been found feeding on the unripe seeds of *Silene nutans* by Mr. H. Moncreaff; these I at once saw were a species of *Dianthæcia* new to me, and, on referring to an extract from the ' Annales de la Société Entomologique de France ' published in 1830, I found there an account of *D. albimacula* by M. Guenée, which seemed to suit them well; I would not, however, venture upon publishing the notes I made of them until their identity had been established beyond doubt, and this has now been done most satisfactorily. Mr. Moncreaff has bred a specimen of *D. albimacula* as early as the 6th of May from the larvæ he collected last summer, by placing some of the pupæ in a warm room. We are now sure, therefore, that *D. albimacula* is a species which breeds in England. For some years it had been relegated to the list of reputed British species (though Stainton's ' Manual ' kept it in its place) until the announcement of its re-discovery in 1865, by the capture of a single specimen which was sent to Dr. Knaggs for identification, as recorded in the 1st volume of the ' Entomologists' Monthly Magazine,' p. 237.

The larvæ I had fed well on the *Silene nutans* that accompanied them, and soon ate out the contents of the capsules, of which Mr. Moncreaff kindly sent a further supply, and when these dried up I found the three younger larvæ (two having already turned to pupæ) take very well both to *Silene inflata* and to *S. maritima*, and between the 14th and 25th of August they retired into the soil prepared for them.

The young larva when a quarter of an inch long is of a greenish-grey colour, and darker than it afterwards becomes ; at this time it has pale dorsal and subdorsal lines, with a darker stripe along the spiracles, bounded above by a paler undulating line ; some faint darker marks along the back indicate the rudiments of the future dorsal design ; a pale stripe runs beneath the spiracles, and the belly is darker greenish-grey. At its next moult, when about three-eighths of an inch long, the ground-colour is either a pale drab or pale ochreous-yellow with the design of dark grey or blackish diamond shapes and spots on the back tolerably distinct ; and when it has attained the length of about three-quarters of an inch the whole pattern of its markings is (as usual) more clearly defined than at any other period, composed as they are of closely aggregated greyish or blackish atoms, which, as the larva grows, become more dispersed with increasing intervals of the ground-colour between them ; but in this clearly defined stage of marking the ground-colour is yellowish-ochreous, the dorsal pattern consisting of a somewhat ovate blackish spot at the beginning, followed by a diamond- or pear-shape extending to the end of each segment ; the front half of each of these pears or diamonds is rather bare of freckles within its outline, showing the ground-colour there more or less, while the hinder part is filled up so as to look blackish ; the anterior pairs of tubercular black dots show distinct on the clear unfreckled ground of the back ; the hinder pairs of dots are often attached to the lateral angles of the diamond-shapes, but not invariably so, though they are always touched by a blackish line of freckles that curves or festoons along from the hinder dot of one segment to the hinder dot of the next; beneath this is the subdorsal interval of clear and paler ground-colour; and then come two broad and irregularly thickened stripes of freckles, which about the middle of each segment slope towards each other till they touch, then returning to their previous level; the ground in the space just below the point of contact is filled with freckles which partly surround the white spiracle outlined with black.

The larva when full grown measures one and a quarter inches in length, is of moderate stoutness, cylindrical, with the head a trifle smaller than the second segment, which is in turn a little less than the third, the anal segment tapering a little behind. Its ground-colour now is pale ochreous or pale brownish-ochreous; the head is delicately freckled and streaked with dark brown down the front of each lobe; the second segment has a dark brown or brownish-grey plate through which the fine dorsal and broader sub-dorsal lines of ground-colour are visible; on the rest the dorsal line can be faintly discerned as a fine thread of ground-colour running through the dorsal blackish spots and ill-defined pear-shapes that follow them; both front and hind pair of black dots are now equally distinct on the back of each segment, and a similar dot is situated a little above each spiracle, which last is whitish, faintly outlined with black; a patch of dark grey or blackish freckles anteriorly in the subdorsal region, and some broken patches of lines of freckles extending in curves to the spiracular region on each segment, are now the only remains of the design mentioned in the previous stage, this change having been brought about by the scattering of the dark atoms which before were confined in lines; the belly and legs are of the ground-colour.

As will be seen from what follows, there is con-siderable resemblance between this larva and some of its congeners, but to my eye *its* most striking characteristic is the *absence* of the *slanting streaks* or *chevrons* which *they* so generally have.

The pupa is little more than five-eighths of an inch long, stout in proportion, the wing-, antennæ-, and trunk-cases projecting in a blunt point over the abdomen, which tapers off gradually ; the abdominal rings are partly granulous ; the colour of the thorax and wing-cases is deep reddish-brown, the abdomen dark brown.

M. Guenée has observed of the larva of *D. albimacula* that " in a manner it resembles that of *D. capsincola*, and when they are together on the same plant they afford fine exercise for the eyes to distinguish them.

" It is found upon *Silene nutans*, and sometimes, but much more rarely, on *Silene inflata*. In captivity it accommodates itself well to these two plants, also to *Lychnis dioica*.

" This caterpillar is not rare where *Silene nutans* grows, that is to say in the arid and hilly places of certain woods." (W. B., May 11th, 1874 ; E.M.M., June, 1874, XI, 16.)

DIANTHÆCIA CÆSIA.

Plate LXXXVII, fig. 3.

Towards the end of June, 1867, I received two small larvæ of this species, scarcely half an inch long, from Mr. C. S. Gregson, who sent with them a couple of flowers and a leaf or two of *Silene maritima*, and instructions to feed them on the flowers and leaves of that plant. Perhaps for want of sufficient air in their small box during the journey in hot weather, one arrived in a dying state, and though much attention was paid to the survivor, that also soon sickened and died.

I noted that this little larva was of a grey colour,

marked with a series of dorsal diamonds of a darker brownish-grey, and the sides of the same brownish-grey, and the tubercular dots distinct as dark rings. The next and only subsequent opportunity for studying this larva was generously afforded me by Mrs. Hutchinson, of Leominster, who, knowing that *D. cæsia* (at that time a comparatively recent addition to our list) was still one of my desiderata, very kindly sent me, on July 23rd, 1869, some examples of larvæ which her son, Mr. Thomas Hutchinson, had recently found on *Silene maritima* in the Isle of Man, in the hope that I might find that species amongst them; nor, indeed, was this a forlorn hope, for on looking over the larvæ—mostly familiar enough—I detected one which at once recalled to mind the *D. cæsia* of 1867. This I kept apart, and carefully tended with seeds of both *Silene maritima* and *S. inflata*, and it appeared to feed very well on both, without any apparent preference. At its arrival it was about five-eighths of an inch in length, by the 31st it had increased to an inch, and by August 9th to one inch and a quarter, perhaps even a little more when thoroughly stretched out; it continued to feed for a day or two longer, and retired into the soil on the 13th for pupation.

Having but this one, I did not interfere with it to take notes of the pupa, lest I might by some accident destroy the chance of breeding the perfect insect, and so lose the opportunity for proving that the figures taken of the larva were rightly named. Fortunately the moth, a fine example, appeared on July 15th, 1870, evidently later than its parents had flown in their native haunts.

In several of the *Dianthæciæ* we are familiar with variations of the chevron pattern, but from all its congeners *D. cæsia* is strikingly distinguished by these diamond-shapes of freckles, as well as by the absence of positive outlines in the subdorsal and spiracular regions. The description given above of the larva of 1867 suits every subsequent stage up to full growth.

The full-grown larva was cylindrical, of moderate stoutness and about equal bulk throughout; the head rounded, and the anal segment only slightly tapered, and rounded off; the segments and sub-divisions very well defined. The ground-colour on the back was pale greyish-ochreous, that of the sides, belly, and legs similar, but a little more ochreous-brown, and deeper; on each segment, and co-extensive with it in length and in breadth, extending to the subdorsal region, was a diamond-shaped mark, composed of dark grey-brown freckles thickly aggregated together; a broad band of similar freckling commenced along the subdorsal region and terminated rather below the spiracles; other freckles, fainter and farther apart, were low on the sides, disappearing gradually towards the legs, which were tipped with broad hooks; the head shining reddish-brown, the plate on the second segment was rather shining reddish grey-brown, and had a broad dark brown margin in front; the tubercular dots were whitish centres in rings of dark grey-brown, arranged in threes on either side the back of each segment; others, whose place was within the side band of freckles, were of the ground colour; the spiracles pale brown, outlined with black; just at the last, when the larva was in its plumpest condition, the last three segments appeared to taper a little.

There was a faint indication at first of a dorsal line or thread of pale ground-colour enclosed within two dark grey-brown ones, but not very visible beyond the thoracic segments, excepting just at the segmental divisions. (W. B., July, 1872; E.M.M., August, 1872, IX, 64.)

DIANTHÆCIA BARRETTII.

Plate LXXXVII, fig. 5.

For the exposition of the habits of this rare species, which, so far as at present known, seems confined to a part of the Irish coast, I am greatly obliged to Mr. E. G. Meek, who kindly sent me nine eggs laid by a captured female, four of them on part of a flower calyx of *Silene maritima*, to which they adhered, and five loose.

I received the eggs in July, 1878, when nothing seemed to be known of the larval food-plant for certain, though I then heard from a kind friend of great experience that *Statice armeria* might be a likely plant to try, as well as that on which the eggs had been laid, and which was naturally also suggested by the insect's generic name of *Dianthæcia*.

Seven of the eggs were hatched in the evening of July 10th, the other two next morning, and the little larvæ were quite remarkable for their activity and robustness as soon as they were out of the shells, marching vigorously over small sprays of the two plants above mentioned provided for them.

During the next day three of the larvæ were eating out little sinuous channels in leaves of the *Silene*, surrounding themselves with frass, and by the third day had worked their way into the stems at the axils of the leaves, where they had also thrown out little heaps of frass. Similar indications showed that one individual had entered a *seed-capsule* from within the flower calyx; the others were still to be seen roaming about at intervals until it occurred to me to try them with a small piece of the root of the plant, as well also of that of the *Statice*, when they all soon after disappeared.

On the fourth day, while inspecting the piece of root of *S. maritima*, I detected two small holes in it with heaps of minute pale cream-coloured frass adhering to them ; and on the seventh day I examined the axil of a stem and leaf, where I found a larva had mined its way downwards and was lying a quarter of an inch below in the stem, waiting apparently for its first moult, but my stripping away half the stem to expose it proved fatal, for it soon after died.

About this time I began to realise the intentions of the infant larvæ, and could but lament the jeopardy my experiment had placed them in, on finding the bits of plant were losing their freshness, and the impossibility of rescuing the tiny creatures from their perilous positions, for each attempt made proved fatal in a short time to those in the stems ; soon, too, the bit of root began to turn mouldy, and a fresh piece was placed beside it, and on the eleventh day I had the satisfaction to see a heap of frass thrown out of it—a proof of one still alive and within the fresh piece.

In the meantime I had satisfied myself that neither flowers, leaves, nor root of *Statice armeria* had been attacked, and therefore had potted two plants of *Silene maritima*, having good roots and close together in the pot, and there, between those roots just beneath the surface of earth, I wedged in tightly the bit of root containing the sole surviving larva, on July 23rd.

On September 13th I turned out the contents of the pot, by inverting it, to search for the larva, but no search was needful, for the soil had been more sandy at the bottom of the pot than elsewhere, and now formed the top of a cone, and this slipping away, the larva at once rolled into view ; it had evidently done feeding, and a great quantity of pale frass, quite fresh, filled up a large cavity in the shoulder of the thickest root ; if it made any chamber or gallery the falling away of the light earth had quite destroyed it.

When the larva was placed in a small pot with some of the earth it did not burrow underneath, but, after much wandering about, eventually settled down under a fragment of root, and there changed to a pupa on September 17th; the moth, a male, emerged on the evening of June 27th, 1879.

The egg in shape is round, somewhat flattened, and with a slight depression beneath, the surface very finely ribbed and reticulated; the colour very pale greenish-buff, and on July 6th, when first in my possession, showed a faintly darker speck at the top, and by the 9th of the month the embryo was seen through the shell as a dark spot on one side, and this next day was increased in size and in depth, of a leaden hue, while the rest of the shell was tinged with brownish-grey, when the hatching soon began.

The young larva at first is of a very pale rather greenish-drab colour, with blackish head and narrow blackish plate on the second segment. After the first moult, when the larva has fairly become an internal feeder, its head is shining reddish-brown, and the narrow shining plate on the second segment is of similar colour but a little paler, the rest of the body tinged with livid reddish-grey, the skin rather shining and semi-transparent, through which a dark internal vessel is visible, the anterior legs reddish-brown.

The full-fed larva measures about one inch and a quarter in length, stoutest at the third and fourth segments, the second tapering a little to the smaller and rather narrow head, with lobes deeply defined at back of the crown, tapering also from the tenth to the anal extremity; the ventral and anal legs short, and well beneath the body; the segments very lightly and finely wrinkled towards the well-cut divisions on the back, the sides much dimpled. The colour of the head is reddish-fawn and shining, the lobes outlined on the face with blackish-brown, defining well the triangular division and the upper lip, and below this the mouth itself. The body is of a light fawn inclining to flesh colour; a narrow scale-like plate, of glossy pale yellowish-fawn colour, is on the second segment, with an interval of the paler skin towards the head; a similar plate is on the anal flap, and a dorsal vessel of brownish-grey shows faintly through the skin; the rather small tubercular dots are fawn colour, each with a short bristle, spiracles black, anterior legs pale fawn colour, the ventral and anal legs with a fringe of dark brown hooks.

The pupa is three-quarters of an inch long, the head and shoulders rounded off, the wing-covers wrapped close to the body, and the antennæ and legs enclosed in a *blunt rounded projection* at their ending, a *little free from the body*; from thence the abdominal rings are deeply cut and taper gradually to the tip, furnished with two small spines. Its colour is dark red-brown until about a week or so before the emergence of the moth, when by degrees paler patches of yellowish-brown appear on the wing-covers; the smooth abdominal divisions are dull, but all the rest of the surface is glossy, although the other parts of the abdomen and thorax are finely punctate.

I have to revert now to that only larva which, whether by mistake or not, ate its way into a seed-capsule, whose appearance in the second stage of its larval life is described in the foregoing. When about to open the capsule I expected to find the larva dead, as the little heap of minute whitish frass made on its first entering had not been accumulating and still remained blocking the small hole, and was hard and dry. But the larva, greatly to my surprise, was alive, had moulted once, had grown and prepared for a second moult, while the unripe seeds were nearly all devoured and converted into frass, perfectly black. On carefully exposing the larva to take note of its altered condition, work of only a few minutes, yet in that short time it became more and more languid, as I judged from the exposure to air, and I hastened to place it inside a fresh calyx with seed capsule, forgetting at the moment that it was unable to use its mandibles, from the head having been too far drawn back from the head-piece in front, in anticipation of moulting; but it soon became inert, and died.

Looking back at the results of my experiment with the eggs of *D. barrettii*, I seem first to have established the fact that it is not a *Dianthæcia*, although it certainly has some affinity to that genus, as shown in the solitary instance of one infant larva out of nine making its way into a seed-capsule and there sustaining itself on the unripe seeds; and again, more particularly, is this shown in the form of the pupa. Next, that from the behaviour of the other eight larvæ they proved clearly enough that their normal habit is to enter the stems of the plant, and through them by degrees arrive at the root, where they feed and mature—a habit well confirmed by the structure and appearance of the larva itself, which, not only when full fed, but even in its earlier age between the first and second moults, agreed so well with Guenée's description of that of *Luperina luteago* (*vide* 'Noctuélites,' tome i, p. 181), that although some disparity of size and colouring exists in the perfect insects, as most obligingly shown to me by Mr. Edwin Birchall, yet I am constrained by the evidence to believe *D. barrettii* to be an isolated and melanic variety of *L. luteago.** (W. B., July 7th, 1879; E.M.M., August, 1879, XVI, 52.)

DIANTHÆCIA IRREGULARIS.

Pl. LXXXVII, fig. 4.

On the 13th of August, 1870, I received, through the kindness of the Rev. E. N. Bloomfield, of Guestling, near Hastings, seven larvæ of this insect. Unfortunately, perhaps from having been too closely packed during the hot weather, four of them were dead on arrival; another one appeared sickly, and soon died, but the remaining two seemed to be quite healthy.

* When last in London, Dr. Staudinger stated to me that, in his opinion, *D. barrettii* is a form of *L. luteago* (R. McLachlan, E.M.M., XVI, 55, foot-note).

They were not full grown, but had, I think, assumed
the markings of the adult larva; were of average
thickness, and about three-quarters of an inch in
length. The head is considerably smaller than the
second segment, and emits a few short hairs. The
body is nearly uniformly cylindrical, but tapering very
slightly anteriorly; the segmental divisions tolerably
conspicuous; the skin smooth and slightly glossy;
the usual dots rather indistinct. The general colour
is pale yellowish-brown, tinged with green; the head
wainscot-brown, sparingly dotted with black. The
dorsal line is composed of a series of narrow V-shaped
smoke-coloured marks, one on each segment, and the
apex of each pointed posteriorly; these V-shaped
marks are most conspicuous from the fifth to the
ninth segments; on the others they appear as an
interrupted greyish line, bordered on each side with
smoke-colour; the subdorsal lines are inconspicuous,
dull white, margined very narrowly with brown;
along the spiracles extends a smoke-coloured stripe,
dark on the lower part, but shading gradually into the
ground-colour above. The ground-colour between
the dorsal and spiracular lines is faintly variegated
with brown, and below the spiracles, which are black,
is a yellowish-white stripe. The belly is semi-translucent and shining, greyish tinged with green.

It feeds in a state of nature on *Silene otites*, but
mine did not refuse *Lychnis flos-cuculi*. In a few days
they began to wander about the breeding-cage, to the
sides of which they finally attached themselves, when
a single ichneumon emerged from each. Each of
these parasites spun its cocoon beside the dying larva;
these cocoons, when completed, were very curious,
and almost exactly resembled dried husk-covered seeds
of some plant. Mr. Bloomfield informs me that a
large proportion of the larvæ of this insect are infested with this parasite. (Geo. T. Porritt, September
2nd, 1870; Entom., October, 1870, V, 177.)

POLIA CHI.

Plate LXXXVIII, fig. 3.

On September 14th, 1869, Mr. Longstaff, then at Forres, kindly sent me a few eggs laid by a female of this species; these began to hatch on April 11th, 1870, and continued at the rate of about one or two per diem until the 22nd. The larvæ, when young, fed chiefly on *Rumex crispus*, and occasionally on some other low plants; they at that time were not particular in their diet, for they seemed to welcome any change given them; but after two or three moults they began to show a decided preference for sallow and osier, and on this food, towards the last, they were entirely kept, until they became mature one after the other, from May 20th to June 5th; the perfect insects appeared from August 10th to 21st.

The egg of *Polia chi* is hemispherical in form, or rather elliptical at top and flattened beneath at its greatest diameter; deeply ribbed and reticulated; when first laid it is yellowish, and in a few days turns greyish-brown, and about a week before hatching a broad zone of flesh colour appears below, while the upper part is a rich crimson-brown; at this time, in respect of colour, variations occur,—some of the eggs have a narrow zone of blackish at a distance of two-thirds from the base, while the top is irregularly blotched with this colour.

The young larvæ, when first hatched, were pale olive-greenish, the large head pale brownish with distinct black dots and hairs; but they became in a couple of days rather bluish-green. By April 23rd the most forward had become half-an-inch in length, very slender, and of a full deep green colour, with the lines and also the dots paler green; by May 15th the biggest had grown to one inch in length, still slender in proportion, and rather less deep in colour, which, by the 19th, had changed to a yellow-green,

and then the fine lines were whitish. By the 27th the
larvæ were in their last coats, and presented but little
variation in details; their attitudes were graceful
amongst the twigs of sallow and willow, as they bent
and elongated themselves when feeding near the
extremities of the leaves.

The full-grown larva is one and five-eighths of an
inch in length, very slender in proportion, cylindrical,
though tapering a little at each end, the head
rounded, and the antennal papillæ well developed, the
segmental divisions very delicately defined, as well as
the intermediate wrinkles, so that the skin appears
very smooth and soft. The colour on the head and
back is a delicate bluish-green, quite opaque on the
back, the dorsal line very thin, a mere pale thread and
edged with darker green than the ground-colour; the
subdorsal line is whitish, better defined, and also
edged with darker, indeed this line on the thoracic
segments is white; beneath it, as far as the spiracles,
the colour is a transparent yellowish, or else a full
green, so clear as to show the branchial vessels
through it; this colour deepens gradually below till it
terminates in a fine blackish-green line, along the
lower edge of which are the white spiracles outlined
with black; a brilliant pure white stripe follows, very
broad along the middle segments, and a little attenuated at each end, often extending along the side of
the head towards the mouth, and ending behind at
the extremity of the posterior leg; all the legs and the
belly are rather paler and more yellowish-green than
the back; after the thoracic segments the tubercular
dots are ranged in threes on either side of the dorsal
line; they are paler than the ground, and sometimes
ringed with a little darker colour.

The variations seem to be in the depth of the
general ground-colour, more or less blueness of the
green, and the presence or absence of a small blackish
oval ring with whitish centre, or false spiracle on the
side of the third and fourth segments.

When full-fed the larva makes a decided cocoon under the surface of the soil, of bits of earth somewhat toughly spun together; the pupa is nearly six-eighths of an inch long, smooth and regular in outline, tapering gently to each end, the last segment of the abdomen terminating rather bluntly and furnished with a knob, from which is emitted a pair of *very fine* (*quite bristle-like*) hooked spines; the colour of the pupa is reddish-brown, and the surface glossy. (W. B., March, 1873; E.M.M., May, 1873, IX, 290.)

POLIA FLAVOCINCTA.

Plate LXXXVIII, fig. 4.

On the 9th of July, 1873, I received from the Rev. B. Smith five larvæ of this species reared from eggs and sleeved on apricot; they also like mint. Four of them were full grown, the other one half grown. The full-grown larva is an inch and three quarters in length, of moderate stoutness throughout, and cylindrical. The head full, and but little less than the second segment in width; the segments very smooth and scarcely defined when stretched out, though on the larva turning itself round the skin folds itself at some of the segmental divisions and then shows distinctly pale yellow in the folds. The ground-colour of one of these is yellowish-green, the others being bluish-green; a slightly darker dorsal line having a line of minute yellowish dots or specks through the middle of it. The spiracular line is dark or blackish-green and is but the mere edging to the yellowish-white subspiracular stripe which melts away gradually below into the green of the belly. The spiracles are white, outlined with black. The ventral surface is green, same as the back, and has a central faint whitish stripe. The legs and ventral and anal legs are unfreckled, but all the rest of the green ground, excepting the head and a smooth velvety

18 POLIA FLAVOCINCTA.

plate on the second segment, is minutely freckled with
atoms of whitish-yellow, a few more distinct than the
rest marking the subdorsal line in an inconspicuous
manner.

The moths emerged on the 21st, 27th, and 28th of
September, and the 1st of October, 1873. (W. B.;
N.B., II, 23.)

POLIA NIGROCINCTA.

Plate LXXXVIII, fig. 5.

On the 12th of July, 1876, Mr. Edwin Birchall,
then at Douglas, Isle of Man, kindly sent me a full-
grown larva of *Polia nigrocincta*, which I received on
the 13th at 3 p.m., and immediately began to figure.
Mr. Birchall informed me that it seemed to prefer
lettuce to its natural food, and was found eating green
seeds of dock, but seeing it had eaten out a flower of
thrift or sea-pink (*Statice armeria*), I provided some
for it, and it ate out several, more or less, during the
ensuing night; some *Plantago maritima* was also
given, both leaves and seeds, but I am not sure that
it ate any, and, indeed, the larva seemed to be almost
or quite full fed, as it measured an inch and three-
quarters in length. It is moderately stout in proportion and cylindrical, of nearly equal bulk throughout,
save that the last segment tapers as usual, and it
tapers also from the third segment to the head. The
head itself is glossy, brownish-red; all the rest of the
body is soft and smooth, without gloss, the segments
plump and well-defined. The colour of the body
above is a peculiar brownish-red, and beneath, the
belly and legs are much of this colour but weakened
by a rather paler mixture of greyish-yellow, so that it is
less red than the back. The somewhat puckered and
inflated subspiracular stripe on the upper margin has
the inconspicuous spiracles; these are of the general
ground-colour most finely outlined with blackish-grey;

the stripe itself is paler than the belly, yet not so very
much, being greyish-ochreous; above the spiracles as
far as the subdorsal region the ground is neutralised and
made rather paler by something of the greyish-yellow
in it; the dorsal is not very distinct, being a thread of
the ground between two freckly lines of dark greyish,
the subdorsal very similar in colour but less distinct.
Over all the ground of the body is a minute reticulation, roundish in character and greyish in colour.
This is also on the under surface as well as the upper;
the abdominal legs are a little darker towards the feet;
the tubercular dots, of the ground-colour simply outlined with darker grey, are in threes each side of the
back of each segment; the reticulation, which has
much the effect of freckles, is rather more distinct
near each end of a segment.

On July 14th the larva, very well behaved and
healthy, ate only a little of sea-pink; on the 15th it
still ate but sparingly, and was evidently decreasing in
length and bulk; on the 16th this continued, and it
became active in roaming and rather irritable when
touched; and by the 17th had retired beneath the
soil. As the moth did not appear, I turned out the
pot on the 6th of October, and found a blackish-brown
pupa with a cocoon of silk covered with grains of
earth, nearly an inch long, of an oval shape, but the
pupa itself was soft and dead. (W. B., October,
1876; N.B., III, 110.)

DASYPOLIA TEMPLI.

Plate LXXXIX, figs. 1—5.

This larva, discovered by Mr. W. R. Jeffrey, has, I
am aware, been described by Mr. Newman, in the
'Zoologist' for 1863, p. 8788; yet, as it scarcely seems
to be reckoned common at present, a few notes on its
earlier stages which I have put together from the

observations of myself and my friends, may not be uninteresting.

In the latter part of the year 1865, Mr. H. Doubleday obtained living impregnated female moths from Mr. Varley, of Huddersfield, and succeeded in keeping them alive through the winter; one even survived a journey to M. Guenée, and, as well as the two retained by Mr. Doubleday, deposited eggs about March 20th, 1866, which he distributed to his friends, Mr. Hellins and myself amongst them. They were laid on the under-sides of some leaves of *Heracleum sphondylium* (then grown out to a foot in length), which were put in a box to induce them to commence laying. Perhaps in a state of nature, when the *Heracleum* leaves are backward, the females may deposit on the dry stems of last year's plants.

The egg is not so flat as the usual *Noctua* shape, but stands up rather higher, is ribbed, at first yellowish in colour, afterwards turning flesh-colour, with a pinkish-brown spot on the top, and a ring rather above the middle; finally turning blackish a day or two before the hatching of the larva.

The larvæ appeared about April 20th; at first they were of a dingy olive colour, with black heads, rather longish-looking in shape.

Mr. Jeffrey having made the entomological world acquainted with the food, we had all provided some *Heracleum* plants ready at hand in our gardens, and put out on them the larvæ immediately upon their appearance; nor had we to wait long in suspense as to their powers of eating. Some began by attacking the leaf itself and afterwards the stem; others made at once for the stem, and commenced eating their way into the interior, and drinking the sap which flowed into their little tunnels; from this point their habits as internal feeders made it difficult to watch their growth, but the following observations were made.

It seemed that on reaching the centre of the stem they proceeded downwards, at first giving no sign of their presence ; but at the end of a month—about May 23rd—they had reached the bottom of the stems and the crown of the roots, and several of the plants began to show signs of decay.   On the 4th of June one larva was extracted from near the bottom of a stem, and by that time measured about five-eighths of an inch. Being placed on another plant it made itself at home there also, and fed and grew till July 10th, when it was again examined, and being found then to measure one inch and three-eighths, was sent to me to be figured.

It appeared very uncomfortable when taken out of its food-stem and exposed to light while being de-picted, and when replaced on the stem soon found its hole and disappeared within.

I attempted to rear it on a cut stem of *Heracleum* inserted in a pot of moist earth, but after a few days it left the stem and died in a very flaccid condi-tion.

Meanwhile the larvæ which had been undisturbed seem to have eaten away and grown more rapidly, penetrating even into the main root of the plants, and causing them to wither.   By July 10th Mr. Doubleday examined one larva which was nearly full-grown, and about this time probably most of them became restless and wandered off in search of fresh food, for about the 18th of the month neither Mr. Hellins nor I could find one left in any of our plants, and Mr. Doubleday had but few remaining.

However, on the 19th he most kindly sent me a large root with two larvæ, then about one inch and five-eighths long, and on the 25th another—a fine fellow, two inches long, and apparently full-fed.

Before describing the larva I may here at once say that neither of us succeeded in rearing an imago; those larvæ that did not run away became infested with parasites (*Microgaster alvearius*), and thus perished miserably.

However, M. de Graslin (to whom Mr. Doubleday

had sent eggs) was more fortunate in France, and
succeeded in rearing several fine moths.

The shape of the larva (after it has attained some
size) is moderately stout, cylindrical, tapering but
little at either extremity, smooth and shining; the
folds and segmental divisions very slightly indented,
a triangular inflation round the spiracles; the tubercular warty spots slightly raised and shining, and all
the legs well developed.

In colour the *half-grown* larva is of a dull flesh tint,
tinged with green beneath and at the segmental
divisions, but much suffused with a deep dull pink on
the back; the warty spots blackish. When *two-thirds
grown* it is wholly of a deep but dull flesh-colour,
slightly suffused on the anterior segments with a
dull red.

The *full-grown* larva is flesh-colour, having the dorsal
pulsating vessel visible as a stripe of a darker tint of the
same. The head is, as in the other stages of growth,
brownish-red, and the mouth dark brown. The reddish
shining plate on the back of the second segment is
divided down the centre by a thin line of flesh-colour,
and is thickly margined in front (where it is widest), and
more delicately behind, with dark brown. On the anal
segment there are four brownish-red plates, thus
placed: on the anterior part above the fold of the anal flap
a central semicircular plate margined with dark brown,
and on each side of it at an obtuse angle a small narrow
plate; the fourth and largest plate is on the anal flap,
and has its anterior edge undulating, and margined
with dark brown, its anal extremity ending in two very
small points. The anterior legs are brownish-red, and
the pro-legs slightly tipped with brown; the oval
spiracles whitish, outlined with blackish. The brown
tubercular spots generally round, but sometimes oval
on the third and fourth segments; on these segments
also they are placed in a transverse row on the back,
and end at the sides in a triangular group of three
larger spots. Altogether there are twelve spots on

each of these thoracic segments. The other segments, to the twelfth inclusive, have the usual two pairs on the back of each, and the thirteenth one pair; the anterior larger than the posterior pair, and all gradually diminishing in size from the fifth to the eleventh segment, but on the twelfth they become larger again, and are there transversely oval in shape.

Each spiracle has a large round spot above and below it, another behind and a small one in front of it; these two last-mentioned are sometimes both small, and sometimes one of them is absent; but the two that are above and below the spiracles are larger than any on the back.

The last larva (sent me on July 25th) I retained, and noticed that before August commenced it had entered the earth to change; but at the end of August or beginning of September, instead of a moth, I observed a myriad of small winged creatures emerging, and, on digging, found a long, conical, whity-brown cocoon, which the little parasitic larvæ had constructed over the remains of their victim. These, as mentioned above, were *Microgaster alvearius*, and it seems a mystery how this parasite can lay its eggs upon the larva of *D. templi*, boring as it does into the leaf-stalks or stems of *Heracleum* as soon as hatched, and the minute orifice it then makes being soon closed by the exudation of sap. Neither does it show itself again (unless the plant fails to supply sufficient food), but in nature gnaws a hole just below the surface of the earth for its escape.

Mr. Doubleday most kindly procured for me two pupa-cases from which the moths had emerged, but which were in excellent condition, so that I could take a good figure of them. The pupa is barely an inch long, strong in texture, moderately stout and uniform in bulk, tapering gradually near the abdominal tip, which is terminated by a thick blunt spike; the rings of the abdomen very plump and deeply divided; the anal spike is black, all the rest of the surface purplish-brown, but shining with a leaden hue, as though covered with plumbago.

I understand M. de Graslin bred his moths in August, but in Yorkshire they are seldom found till the third week in September. (W. B., 1868; E.M.M., April, 1868, IV, 251).

Epunda lutulenta.

Plate XC, fig. 1.

On the 8th of October, 1868, Mr. Henry Terry, of St. Marychurch, captured a female of this species, and having induced her to deposit her eggs in captivity, he kindly sent me a portion of them, retaining some for himself and sending others to the Rev. E. Horton.

The egg is circular, a little depressed at the top, and flattened beneath, ribbed and beaded; when first laid it is of a canary yellow, and changes in a few days to a pale pinkish grey-brown, having the top and a broad zone round the middle of the sides of a much darker tint of the same; in about a month it changes to a purplish-grey tint, and just before hatching assumes the bloom-like appearance of a purple grape.

This last change was simultaneously assumed by all the eggs in my possession on November 22nd, that is about six weeks after they had been laid, but from some reason or other unknown to me, no more than two larvæ were hatched out; my friends, as will be seen below, were more fortunate.

The young larva at first has a very dark purplish-brown head, the body pale dirty-greenish and trans-lucent; the internal organs showing through the skin give the appearance of a broad dark grey stripe down the back; there is a dark brown plate on the second and on the anal segments; the tubercular dots distinct, and blackish, each having a rather long dark brown hair.

My young larvæ fed freely on *Poa annua,* but the

grass becoming infected with mildew, they both suddenly died on the 14th January, 1869. I am, however,
able to carry on their history, Mr. Terry having kindly
forwarded me some of his batch on February 20th; these
were then three-eighths of an inch long, of a full green
on the back and sides, the ventral surface rather paler.
The most noticeable feature at that time was the subspiracular stripe being whitish or greenish-white in
some, yellowish or of a pale flesh tint in others ; and by
the aid of a lens one could see that the dorsal line was of
the ground-colour, finely outlined with darker green,
and the subdorsal paler green also outlined with darker;
also that the ground-colour of the back was delicately
freckled over with darker green, the head and an unfreckled plate of green on the second segment, both
paler.

These individuals fed tolerably well for some days
on mixed grasses sown in a pot, and they varied their
food a little by feeding on some of a miscellaneous
collection of plants that had sprung up with the grass,
especially on *Potentilla fragariastrum*, leaving chickweed and trefoil almost untouched ; however, they
had never seemed healthy since their arrival, and they
soon began to die off, the longest-lived going about
the middle of March.

Soon after this, I became aware that the Rev. E.
Horton had been more successful, and though his
stock of young larvæ kept out of doors during the
winter had been a temptation to robins as choice
morsels of food not to be resisted, yet there remained
one solitary individual uneaten, which he most kindly
entrusted to my care, and on May 8th I had the satisfaction of figuring it.

This larva was then one inch and one-eighth in
length, and moderately stout, of the usual *Noctua* form;
its colour a bright yellowish-green, finely freckled with
paler green, the segmental folds showing yellow ; the
dorsal stripe of darker green, the subdorsal stripes of
very pale rather dull yellowish-green; the spiracles

whitish placed on a thin dingy red line, and close
beneath them a rather broad stripe tapering at each
end, of greenish ochreous, edged above and below with
whitish (the whitish edgings of this stripe appear to
me to be the most distinctive mark of the species) ; the
ventral surface and legs of the same colour as the back.

On the 19th of May, Mr. H. Terry succeeded in
finding a nearly full-grown larva on grass in its native
haunts, and subsequently two or three others on
flowers of wild mint and the leaves of *Scabiosa
arvensis*; these he also forwarded to me; they were
then an inch and a half in length, rather darker and
less brilliant in colour than the one reared by Mr.
Horton, but otherwise similar, even in the details,
with the exception that the spiracles were pinkish-flesh
colour, delicately edged with black, and each situated
in a purplish-red crescentic blotch; the plate on the
second segment slightly tinged with the same colour,
and in the middle of the subspiracular stripe there
was a streak of dull pink beneath each spiracle.

On June 19th, I received another larva from Dr. F.
Buchanan White, who had found it feeding on heather
in Inverness-shire; this would not touch grass, but fed
up on heather within a few days after I had it.

This larva was of the same form and character as
the foregoing, though the ground-colour was a rather
bright olive-green, and the dorsal stripe becoming
suddenly blackish on the fifth segment continues so to
the twelfth, being intensely black just at the beginning
of each of these segments; on each of the same segments there was a black streak anteriorly on the upper
edge of the subdorsal stripe; there was also a fine
black spiracular line, interrupted only by the spiracles
themselves and at the segmental divisions.

Although, as I said, this last-named Scottish larva
refused grass, yet from what I could see of the others
I am of opinion that this species is a veritable grass
feeder, probably eating grass all through any mild
weather that may occur in winter, and in spring

probably attacking any low plants that may suit its taste. It seems also that the larvæ invariably both feed and rest on the blades of grass with their heads downwards. (W. B., January, 1870 ; E.M.M., March, 1870, VI, 235.)

Eggs were laid between the 18th and 22nd of September, 1883, by females taken in Darenth Wood, Kent, by Mr. B. A. Bower, from whom I received the eggs on the 28th of that month.

The egg is pinkish-brown at the top with a whitish zone, that is to say the top of the egg in the centre is pinkish-brown speckled with darker brown and surrounded by a broad whitish ring having at its lower ragged edge a ring of dark brown, and beneath this the colour is light pinkish-brown. The general shape is rounded and a little flattened below, and the shell is ribbed and reticulated. On the 1st of October the dark parts of some of the eggs turned darker purplish-brown, which made the broad white ring more conspicuous, but by the 12th this ring had gradually become of a drab colour, hardly noticeable from the general colour, which had changed to grey-brown on the rest of the shell. On the 21st of October I moistened with water some of the eggs, and next morning I found eight of them had hatched. (W. B., October, 1883 ; N. B., IV, 215.)

EPUNDA NIGRA.

Plate XC, fig. 2.

Description of four varieties of the larva.—I am indebted to the kindness of the Rev. John Hellins for many examples of the very beautiful larva of this species, collected in the spring of 1866 by Mr. Thomas Terry and others, found chiefly on *Galium mollugo* and other low plants, though in confinement they preferred hawthorn.

When full-grown they attained from an inch and a

half to an inch and three-quarters in length, and were
cylindrical, of nearly uniform thickness, with the head
rounded, and but little smaller than the second segment, the anal segment tapering at the extremity.

They were full-fed from the middle of May to the
10th of June, and the moths emerged towards the end
of October.

There are several very distinct varieties, and others
that may be subdivided into further variations, but it
will be sufficient to give details of four, distinct in
colour.

First variety.—Ground-colour a brilliant pale yellow-
green, sometimes a very bright grass-green ; others of
a bright olive-green, deepest in tint at the extremities,
and often suffused with pink anteriorly.

The dorsal broad stripe in some being faintly
blackish, in others reddish, but intensely black or
red, forming a dark mark, just at the beginning of
each segment; in others this dark spot is confluent
with the two anterior tubercular large black dots,
thus forming blunt arrow-head marks pointing forwards.   The subdorsal stripe of red or blackish is
sometimes complete, but oftener interrupted in the
middle of each segment; the skin-folds at the incisions are bright yellow.

The spiracles in all varieties are white, placed in
semicircular black marks, and immediately beneath
them is a narrow stripe of pale sulphur-yellow or of
greenish-yellow ; belly and legs pale green, tipped
with red at their extremities. The tubercular dots
are sometimes absent, but when visible are black, and
the anterior pair very much larger than the pair
behind them.   The head green, suffused with pink ;
a dull pinkish plate on the second segment.   One
olive-green larva had the plate dark red, and a large
crimson spot on the top of each lobe of the head.

Second variety.—Ground-colour cinnamon-brown ;
a narrow pale greenish-yellow stripe beneath the
spiracles ; the folds of skin at the segmental divisions

greyish; dorsal stripe faintly indicated by a dusky spot at the beginning of each segment; the subdorsal stripe more distinct and faintly blackish.

Third variety.—Deep dingy crimson on the back and sides; below the spiracles a greenish-yellow stripe; the belly and legs, with head and dull plate on second segment, rather paler than the back; a broad dorsal and narrow subdorsal stripe of faint blackish, but just at the beginning of each segment quite black.

Fourth variety.—The whole of the back between the subdorsal lines a brilliant deep citron-yellow, the sides from the subdorsal to the line of spiracles of the same ground-colour, but almost entirely suffused with dark red; the head and thoracic segments, with the anal extremity, also suffused with red. The dorsal stripe composed of two red confluent lines forming a broad stripe, with blunt arrow-head shapes of red at the beginning and end of each segment, and anteriorly margined with short black streaks; the tubercular dots black, the anterior pairs being much the largest; subdorsal line black, and interrupted in the middle of each segment.

Spiracles white in semilunar blotches of black, and edged below by a pale greenish-yellow stripe; belly greenish, with a large red blotch along the sides above the legs, the latter being orange-red. (W. B., 1867; E.M.M., September, 1867, IV, 87.)

On the 13th of October, 1868, I received from Dr. F. B. White, then at Achilty, by Dingwall, Ross-shire, some eggs of this species. They were laid in a chip-box.

The egg is rounded but flattened at the base, the top having a slight approach to being conical. It is numerously ribbed and beaded, in colour pinkish-brown and very shining.

By the 15th of November they had become darker in colour, and began to hatch on the 21st, and the young larvæ were all out on the 22nd.

At first the young larvæ were of a pale dirty green,

with brownish heads, and after their first moult
became pale yellowish-green with brown heads ; tubercular dots dark brown, each having a fine hair.

They soon began to show a decided partiality for
grass over the various low plants supplied to them,
and after their second moult they became rather pale
and bright yellow-green, rather long in proportion to
their thickness, and by the middle of December they
retired beneath the axils of the stems of grass. From
the 24th to the 30th of that month the grass was
attacked with mould, which killed them off. (W. B.,
January, 1869 ; N.B., II, 144.)

AGRIOPIS APRILINA.

Plate XCI, fig. 2.

On the 17th of November, 1883, Messrs. J.
and W. Davis sent me a dozen eggs, laid together
on chip side by side irregularly in a group of ten
and another of two. The shape of the egg is round,
slightly subconical or depressed dome figure, having
about fifteen stout ribs of pale greenish-drab, the
interspaces black and also the central top of the shell.
(W. B., November, 1883 ; N.B., IV, 210.)

PHLOGOPHORA METICULOSA.

Plate XCI, fig. 3.

On the 18th of September, 1881, after gathering
food for other larvæ, I found three *Noctua* eggs, two
of them laid on a leaf of *Stachys sylvatica,* one being
on the upper side and the other on the under side of
the leaf, and the third egg on the upper side of another
leaf of that plant.

The egg is of a good size, hemispherical in shape,
boldly ribbed from a circular ring or flattened boss
at the top, with shorter ribs filling the spaces
between the long ones, and plainly reticulated. The

colour when found was light straw, and the egg was marked with two faint crenulated zones of pinkish, one below the other, the surface glistening. On the 24th the colour of the eggs changed to a cloudy greyish tint, and on the 25th to a leaden hue, and in the afternoon they hatched.

The young larvæ marched about vigorously as semi-loopers with twelve legs; though the two other anterior pairs of ventral legs are present, they are yet so small as only just to be detected. The colour of the larva, including the head, is lightish green, with a darker green internal vessel showing through the skin, which is dotted with black, each dot bearing a black short bristly hair. In four days the alimentary canal became very deep green in parts, though elsewhere the skin is translucent and almost colourless, and the head pale whity-brown. The larva when disturbed coils the head round over its back, and at this period eats small holes through the leaf of *Stachys sylvatica*.

Their first moult was from October 3rd to 5th, when they were rather less translucent, and their bodies of a subdued dirty greenish tint but still shining a little, the black tubercular dots smaller in proportion and less noticeable, and the two anterior pairs of ventral legs more visible. By the 16th their bodies had become darker, of a semi-transparent green, with a decidedly pale dirty yellowish whitish opaque spiracular stripe, and the twelfth segment so rotund as to create my belief of their being *Ph. meticulosa*. On this day one was laid up, apparently for moulting, as its body seemed empty, being such an uniformly pale yellow-green colour; late in the evening the other two also were laid up, but their darker colouring not yet changed until the next day, when they also were of a light yellow-green colour.

On the 21st of October all three had moulted the second time. I now identified them for certain as *Phlogophora meticulosa*. They were now of a darkish

velvety green, showing the characteristic whitish fine
broken dorsal line which appears only just at the beginning and end of each segment. The small tubercular
dots are whitish, and a pale whitish spiracular stripe
attenuated at each end extends down the side of the
anal legs. The subdorsal line is a rather darker green
than that of the back, and has a line of whitish faint
dots or freckles along its lower edge. The head is
pale green.

On the 2nd of November they laid up for moulting,
and the third moult occurred on the 5th and 6th.
They were now decidedly velvety-green, with whitish
broken fine dorsal line and tubercular dots, and pale
spiracular whitish green stripe edged above with dark
green; the subdorsal dotted line, faintly paler than
the ground, is edged above with a dark green slanting
line nearly as far as the second dot from the front
of each segment. The head is light green, streaked
down the middle of each lobe and reticulated with
brownish-green ; length directly after moulting a little
more than five-eighths of an inch, increased by five
days' feeding to seven-eighths, and stout in proportion,
and in seven days more they were fully an inch long.

They moulted the fourth time on the 20th, 21st, and
22nd, and were during the first day of a much darker
velvety-green colour, the head streaked and reticulated
with brown on the green ground. The very fine
broken dorsal line shows quite white, and also the
very minute tubercular dots surrounded partly with
dark brownish-green. A dark green diamond-shape
appears on each segment beyond the thoracic, and from
their side angles a dark brownish-green streak proceeds
obliquely forwards and slightly downwards through
the next segment in front, fading away just before
reaching the dark edge of the pale spiracular stripe,
which is palest down the side of the anal leg and finely
edged above with dark green along the body. The head
is brilliantly polished, semi-transparent green in colour,
streaked down each lobe and reticulated with brown,

affording great contrast of texture with the soft velvety skin of the body.

They continued to feed well on *Stachys* until it failed on the 7th of December, when leaves of *Rumex pulcher* and *R. crispus* were substituted, and eaten by the larvæ with avidity, and the fact of the fourth moult being their last was soon evident by their quickly attaining the length of one and three-quarter inches when stretched out, being asleep and hiding between the leaves during daylight and coming out to feed only at night; by the 15th they began to mature and were very fat, and the biggest laid up on the 23rd to purge itself of its grossness and fed no more, and becoming reduced in bulk entered the earth on the 27th, the spiracular stripe having faded almost entirely away. The remaining two larvæ underwent similar fading and decrease of bulk, feeding however a few days longer, and they entered the earth respectively on the 30th and 31st of December.

Recapitulation and Summary.

Three larvæ from eggs hatched September 25th.

First moult, October 3rd to 5th	=	about 8 days.
Second ,, October 21st	=	,, 16 ,,
Third ,, November 5th to 6th	=	,, 16 ,,
Fourth ,, November 20th, 21st, 22nd	=	,, 16 ,,
Spun up in earth, December 27th, 30th, 31st	=	,, 35–39 days.

In a cocoon one and a quarter inches long by seven-eighths of an inch wide, of elliptical shape, a little below the surface, one having a leaf adhering, the particles of earth not very strongly adhering to the cocoon, which is of weak texture. The pupa with two anal points surrounded at the base with a few curly-tipped bristles. (W. B., Jan., 1882; N.B., IV, 102.)

34   PHLOGOPHORA EMPYREA.

PHLOGOPHORA EMPYREA.

Plate XCI, fig. 4.

On the 24th of February, 1874, I received from Mrs.
Hutchinson, then at Douglas, Isle of Man, four larvæ,
which she had reared from eggs sent to her by Mr.
Jenner from Lewes, Sussex.

The larvæ on arrival were about five-eighths to seven-
eighths of an inch in length, and feeding on *Ranunculus repens.* They were all of a pale yellowish-green
with one exception—a rather bluish-green,—but all
rather darker green at each end. On the 21st of
March they had grown to from one to one and one-
eighth inch in length, and were very yellow-green,
except just at each end, giving me the fear of their
being sickly, though up to the present they fed, but
were very sleepy and torpid, and now again seemed to
prefer *R. repens,* rather neglecting *R. ficaria,* both of
which plants were growing in their pot.

The dorsal and subdorsal lines, mere threads, are
rather whitish margined with darkish green, and above
the spiracles is an exceedingly fine hair-like line of
whitish; the spiracles are very minute, of whitish
faintly outlined with black; the tubercular dots, small
and whitish, are in threes on either side the back, and
on the side, and one on the hair-like line a little behind
each spiracle; on each segment a very fine silky hair is
emitted by each dot, and several from the head, which
is smallish, pale green and shining. A smooth dull
plate is on the second segment. The ground-colour
of the body is very finely freckled with a deeper green;
the belly is the same as the back; *all* the legs were green,
tipped with brown hooks.

On the 24th of March I received from Mrs. Hutchinson a dark brown example from the same brood, an inch
and a quarter in length, which had originally been
pale green. Its ground was a faintly greenish-tinged

ochreous, but much freckled or reticulated with dark
brown, especially along the sides ; the belly being
a greenish-grey without freckles, and also the ventral
and anterior legs, the head pale brown, faintly reticulated with a little deeper brown, the velvety dark
brown plate showing faintly the dorsal and subdorsal
lines a little paler, and bearing two pale dots on
each side the dorsal line. Beyond the thoracic segments there is on each of the others a dorsal diamond
composed of dark brown freckles thickly aggregated
together, and on the edge of this dark brown diamond
are the small white tubercular dots or trapezoidals, and
the other dot on each side near the beginning of each
segment, for they are in threes each side of the back,
though the trapezoidals are most conspicuous, as they
are either backed or ringed with black. The thoracic
marks of dark brown are somewhat narrow and pear-
shaped ; the subdorsal line is faintly visible, and
though very fine is seen to be edged with fine freckled
dark lines ; the space between this and the equally fine
or finer spiracular line is very thickly freckled with dark
brown. The spiracles are flesh-colour, delicately edged
with black, and are accompanied by a white tubercular
dot on each side and one above, in a blackish-brown
spot ; below the spiracles the surface is thickly covered
with dark brown freckles. The thoracic segments and
the three posterior ones are much suffused with dark
brown.

It should be noted that each dorsal dark diamond
has a paler less freckled spot within on either side the
dorsal line.

On the 1st of April I noticed one of the green larvæ
resting belly uppermost on a rather open network of silk,
which it had spun beneath the upper leaves of *Ranunculus repens*; here it has remained for three days.
(W. B., April, 1874 ; N.B., II, 55.)

APLECTA HERBIDA.

Plate XCII, fig. 1.

On the 6th of August, 1875, Mr. J. G. Ross kindly
sent me five young larvæ of this species feeding on
dock; they were about three-eighths of an inch long, of
stoutish figure, and ate but sparingly for some time,
and in a few days one of them died. Whereupon I
changed their food to *Plantago major*, still giving
dock with it a few times until their decided refusal to
touch it while the plantain was present induced me
to confine them entirely to this latter food, with the
hope of feeding them up by the approach of winter.
Their progress was very satisfactory. They were
half-grown by the 12th of October, and on the 18th
of November one, full-grown, burrowed into the earth
to spin up, another, full-grown, seemed ready to
follow, and two others were not very far behind.

When first they arrived they had most of their
distinctive markings very plainly developed on a paler
ochreous ground than it afterwards became, and their
brown-coloured dorsal marks and freckles had something of greenish in the brown, but as they grew they
became more decidedly brown, the four varying but
little in the depth of their colouring, and on attaining
their half growth they were all very uniform in
appearance. Their peculiar colouring reminded me
of the snuff-box of a friend containing brown rappee
of just their colour. At this stage one failed not to
see three faintly paler lines beginning on the thoracic
segments, but only the central or dorsal one continuing
down the rest of the back, and intersected at the end
of the twelfth segment by a transverse pale margin
terminating abruptly, or rather obscure dorsal diamond and side wedges thus truncated; on the other
segments a dorsal complete diamond is just visible
of brown freckles very little deeper in tint than the
ground, which from their youth has been steadily

increasing its depth; along the side is a similar series
of more elongate and narrower diamonds of freckles,
followed below by a rather thin blackish line on which
the white spiracles delicately edged with black are
seen.   Beneath the spiracles is the usual stripe of pale
ochreous.   On the head is a black brown streak down
each lobe, and the spaces between the lines on the
second segment are filled with rather darker brown
than the rest of the back.   The lateral breadth of the
junction of the twelfth and thirteenth segments is the
widest part of the larva ; otherwise, with the exception
of the head and second segment being a little tapered,
the rest of the body is cylindrical and of uniform sub-
stance.

These larvæ fed up to full growth in November, two
of them burrowing into the earth on the 17th, and the
other two on the [no date given in MS.].

The full-grown larva and from its last moult is very
much darker brown, indeed for a few days after this
moult it is so dark a purplish-brown as to show but
little more than the dorsal paler fine line, but with its
growth by expansion of skin the slightly paler subdorsal fine line also appears, but in an interrupted
manner.   The ground-colour of the skin is now of a
pinkish grey-brown or light purplish-brown ; this is also
the ground-colour of the very shining head, which is
broadly streaked with black down the front of each
lobe and delicately reticulated at the sides, with a black
triangle between the lobes.   The second segment has
a velvet-like blackish-brown patch on the back through
which run the fine dorsal and subdorsal lines ; the
last-mentioned is on the next four or five segments
invisible from the dense mass of dark blackish purplish-
brown freckles that hide the ground-colour, but this
begins to appear on the other segments more and
more distinctly towards the posterior end of the larva,
and it is along those segments where the subdorsal
line is best to be seen, and within this line on either
side is a blackish purplish-brown wedge mark com-

posed of closely aggregated freckles and uniting much
with a central somewhat urn-shaped form of dark
similar freckles. These dorsal markings leave the
paler ground-colour both at the front and hind
margin of each segment comparatively clear, so that
each dark freckled shape assumes rather a squarish
character on the eleventh and twelfth segments,
an urn shape on the others; a fine blackish spiracular line not much interrupted, bears just above
it, anteriorly, a blackish triangular blotch, and close
to the white spiracle on the blackish line above
mentioned which separates the dark freckly side
from the rather paler ground. Colour of all beneath,
including the belly and legs, of light pinkish-brown.
The subspiracular stripe shows almost a whitish upper
edge on the thoracic segments, but for the rest can
only be distinguished by its inflation, there being no
change of tint; this is sparingly sprinkled with dark
freckles, which are continued below and as far as the
ventral legs; these last are unfreckled; a yellowish
whitish streak runs down the front of the anal leg;
the belly is deepest in tint on the anterior segments,
with a transverse series of dots on the fifth and sixth,
besides freckles on all of these towards the head; the
ventral feet with brown hooks. The tubercular dots
are blackish, raised but most minute, sometimes partly
ringed with the paler skin but only to be noticed with
a lens.

This larva is when full-grown an inch and five-
eighths long, and stout in proportion, the stoutest
segments being the twelfth and eleventh; from thence
to the fourth the thickness is uniform and cylindrical;
the third tapers towards the head, which is very little
retractile within the short second segment; the segmental divisions and subdividing wrinkles on the back
and dimples on the sides are well defined. The
bristly hair from each tubercular dot is very fine and
unnoticeable.

The last larva lay some weeks in a torpid state, and

its change to a pupa was very gradual, but completed by the 1st of February, 1876. Mrs. Hutchinson sent me an example of some she had reared chiefly on strawberry leaves; they were all in their last coats by the 17th of February, 1876, and similar to mine above described.

This last pupa lay in a hollow of the earth, covered only by a leaf, and had the larval skin sticking to the tail and beneath the abdomen. It is a little more than three-quarters of an inch long by a quarter of an inch thick at the widest part, decidedly a stout pupa. The wing-covers rather short, otherwise much of the ordinary form, very dark purplish-brown and shining.

On the 19th of June a crippled male *A. herbida* appeared from another fine pupa enclosed in a thin and extremely brittle cocoon of earth, though smooth inside yet with very little silk; in this instance the old larval skin also adhered to the tail of the pupa, but when removed the tail was found to end with two curled-topped spines a little diverging; the wing-covers of ordinary length and development. (W. B., June, 1876; N.B., III, 54.)

Aplecta occulta.

Plate XCII, fig. 2.

For a complete set of figures of the larva, and the opportunity of studying the history of this species from the egg, I have been indebted to many kind friends, beginning with Dr. F. Buchanan White, who, on October 6th, 1868, sent me four young larvæ swept from heather at Achilty, Ross-shire, which, though put on a growing plant, died in the following February. Next I received on May 1st, 1869, from Mrs. Hutchinson, a full-grown larva brought safely through the perils of hibernation, but which unfortunately died soon after, while in the process of changing to pupa In the same year, on the 18th of August, Mr. Longstaff, then staying at Cluny Hill in Moray-

shire, forwarded me part of a batch of eggs laid
altogether in a heap by a female moth he had imprisoned for the purpose; the eggs were laid two and
even three deep in parts of the heap ; they hatched on
the 27th and 28th of the month, and the larvæ were
reared, some to full growth, by the end of October,
pupating in November, and others again at the end of
January, 1870 ; the remainder of the brood continued
to look well until the end of February, when a death
or two occurred, and through March they died off
rapidly, the last dying during the first week of April,
when about one-third grown ; a fatality also attended
the pupæ, as no imago resulted from them.

The attainment of the final metamorphosis, com-
pleting the history of *A. occulta*, I owe to the kindness
of Mr. J. B. Blackburn, who on his return from
Rannoch presented me, on August 29th, 1874, with
twenty young larvæ, then between two and three
weeks old, which he had reared from eggs laid by a
very black female captured there. Some of these
soon outstripped their companions in growth, the
earliest changing to a pupa on September 22nd, and
others at intervals up to December 4th; and from
some of these four moths were bred on October 13th,
November 23rd, December 7th and 22nd, respectively,
four pupæ still remaining.

Of the larvæ that continued to hibernate quite
small up to the middle of March, 1875, I have been
unable to save any ; for after moulting twice they
seemed too weak to feed, and died mere empty skins,
the last on the 6th of April about three-fourths grown.

The food on which Mrs. Hutchinson reared her
larva was heather, bramble, sallow, and *Rumex crispus ;*
and to those reared from eggs I at first gave *Polygonum aviculare*, though their first meal was on the
egg-shells, which they totally devoured; afterwards
they had, besides the *Polygonum*, sallow and heather,
birch and bramble, *Vinca major* and *Rumex pulcher*,
and the last larvæ from Mr. Blackburn were fed on

Polygonum, then on birch and sallow, with bilberry, afterwards dock and bramble, finally on dock, sallow buds and catkins.

The egg of *A. occulta* is globular in shape, a little depressed on the summit, and rather flattened beneath; the shell ribbed and finely reticulated, of a pale straw colour when first laid, afterwards becoming a pinkish-drab, and at last a dark lead colour.

The newly-hatched larva is of a pellucid whitish-green, with minute black dots; on the third day becoming greener on the back, yellow on the sides, and head pale brown; after moulting twice it is yellowish-grey on the back and belly, dark greyish-brown on the sides, the dorsal and subdorsal lines dirty whitish, the latter edged above with black near the end of each segment to halfway along the next; the pale yellowish-white subspiracular stripe, so characteristic of this larva, now first appears with a black line above it. On becoming five-eighths of an inch long it is so dark as to appear almost black, though in reality the sides are darker than the back, especially towards the spiracles, where the blackest part being in contrast with the pale yellowish-white stripe below, makes it appear very brilliant; the black subdorsal streaks have now become thickened into wedge-shapes, broadest at the twelfth segment, where their bases are only separated by the thin and much interrupted yellowish-grey dorsal line. At this stage in their captivity it was that the precocious individuals began rapidly to increase in size, and attain full growth in autumn, some of them keeping almost black to the last, others showing a mouse-coloured ground-tint, more or less, between the black markings; in these lighter examples the black marks were greatly reduced, in two instances to the merest rudiments.

The full-grown larva measures nearly two inches in length, is stout in proportion, cylindrical in figure, though tapering a little at the thoracic segments to the head, which is the smallest segment; the thirteenth,

sloping down from the back, tapers a little towards
the end, which is rounded off. The general appearance is plump and full, though the segmental divisions
are very well defined, and the two usual transverse
wrinkles towards the end of each segment can, in their
plumpest state, be generally seen.

As regards colour, the head is usually brownish-grey,
streaked on the front margin of each lobe, and reticulated at the sides, and freckled above the mouth, with
blackish or with dark grey; the ground-colour of the
body varies in individuals from a light mouse-colour
to the deepest greyish-brown; on the second segment
is a semicircular smooth but dull plate of rich reddish-brown, edged with black in front, through which
run the beginnings of the usual lines, which are also
continued faintly through the similar brown-coloured
anal flap; the dorsal fine line is in most cases pale
yellow, sometimes, at the very last stage, seen quite
uninterrupted, but often much obscured; the yellow
subdorsal line, a trifle thicker, runs its course in a fes-
tooned manner, when visible, forming a series of
curves, the end of each curve bearing the hinder tubercular yellow dot; the dots, in threes on either side of
the back of each segment, are always visible, and sometimes dingy ochreous-yellow, but the subdorsal line
is sometimes absent; within the subdorsal line on the
back of each segment, in front, is a more or less broad,
black, velvety, blunt wedge-shaped mark, and the
ground-colour between these marks often so thickly
covered with blackish coarse freckles as to give a
blackish appearance to the whole area of the back; on
the side, as far as the spiracles, the ground-colour is
often quite as much obscured with black freckles, while
in some examples this part is freckled equally with
yellow and black; but it is always bounded below by
a velvety-black fusiform or triangular mark bearing
just within its lower edge the black spiracle, which,
though not readily seen, may often be observed to be
delicately margined with grey; immediately beneath

comes the broad stripe of conspicuously bright yellowish-white, narrower on the second segment, and widening gradually to the fifth, suffused in the middle
of each wth a tinge of orange or of pink, and having
a chain-like series of blackish and grey freckles running through its middle ; the belly and legs of the
ground-colour are generally much paler than the back,
but freckled with black at the sides, more sparingly
towards the middle.

Among the larvæ sent me by Mr. Blackburn some
beautiful varieties were developed. Directly after their
last moult they seemed to be quite black and velvety,
but with a brilliant subspiracular whitish stripe ; as the
skin became more expanded by their increasing growth,
the ground-colour began to appear by degrees on the
back and sides, in the interstices of the black freckles, of
a cool grey tinged with a *rosy hue*, and banded across
the front of the segments with a suffusion of blackish-
brown. Those that hibernated and moulted in the
spring, and attained to half and three parts growth,
were differently coloured from any of the others, for,
although the details of markings were similar, the
ground-colours were rich, warm, bronzy-browns.

The pupa of *A. occulta* is nearly one inch in length,
stout in proportion, of the usual *Noctua* form, the tail
ending in two small points a little divergent ; the
surface roughened by minute pits and striations,
except at the divisions of the abdominal rings. In
the newly-changed pupa these were flesh-colour, but
after a few days became dark red, and soon after
turned like the colour of the rest of the surface, a
blackish-purple ; through these parts the wings and
antennæ cases still have the purple rather redder than
the rest ; the spiracles blackish. (W. B., 15th June,
1875 ; E.M.M., August, 1875, XII, 66.)

APLECTA TINCTA.

Plate XCII, fig. 4.

On the 9th of September 1874, three larvæ that had been reared from the egg by Mr. Charles J. Buckmaster were sent to me with an account of their previous history as follows :

"The eggs, about twenty in number, were laid by two moths (taken at sugar at Rannoch) on the 2nd and 3rd of August. They agree with *A. occulta* in shape, sculpture, and markings, but differ in size, being much larger, and in their numbers being laid in loose batches of five or six together, several of them being deposited singly, and in what seems a curious fact, that unlike *A. occulta* and *A. nebulosa*, they are covered loosely with hairs ; this applies both to the small batches and to the single eggs. Out of the total number of twenty-three, seventeen hatched on the 12th and 13th of August ; the rest did not change from their original whitish hue, and eventually shrivelled up. I have taken no description of the larva, which is now in its third moult, but find noted in my diary that on its first appearance it is much darker and larger than *A. occulta* at a similar period ; also that the tubercular dots are much more conspicuous. Its staple food has been *Polygonum aviculare*, accompanied occasionally by birch and *Plantago lanceolata*, for neither of which last was much enthusiasm manifested."

On arrival these larvæ measured three-quarters of an inch, one of them a little more, in length, and were rather slender in proportion. Their ground-colour was light orange-brown, thickly sprinkled with freckles of whitish flesh-colour ; the dorsal marking is a fine double line of blackish enclosing a line of whitish broken up into dots of irregular lengths ; the subdorsal line is similar but less conspicuous from the confused mass of both blackish and pale freckles covering the sides as far as the spiracles, yet a zig-zag darker series can just be discerned ; the head is cinnamon-brown, streaked on each lobe and reticulated with dark brown ; the tubercular whitish dots on the back are margined with blackish-brown only on their side nearest the dorsal line, while the dot above each spiracle appears equally ringed with dark brown. Each dot bears a fine blackish short hair which has the appearance of a black centre ; the spiracles are small and flesh-colour outlined with black ; below them the sides are freckled with pale flesh-colour, the belly and legs of the ground-colour, unfreckled. In the largest of the three a faint appearance of dorsal diamonds was suggested by the aggregation of dark freckles about the dorsal line towards the end of each segment.

On the 15th they had moulted, and by the 29th were an inch in length, their general appearance much as before, their skin having a very mottled look, the light warm-brown ground-colour being freckled both with flesh-colour and deeper brown, in addition to the blackish freckles about the dorsal whitish freckled line ; the black rings to the dots and the sides present to view a still more mottled appearance.

On the 18th of February, 1875, two of them were found dead and stiff, and on the 14th of March the last one was found dead, having some days previously eaten very sparingly of bramble and grass ; it had for some time been getting smaller. (W. B., March, 1875 ; N.B., II, 121.)

APLECTA ADVENA.

Plate XCII, fig. 5.

At the beginning of July, 1865, Mr. Doubleday kindly sent me several young larvæ of this species which had been reared from eggs. They were very lively creatures, greatly averse to light, and very active in crawling and burrowing under their food for shelter. They throve well on *Polygonum aviculare*, and by the 26th of July were three-quarters of an inch in length ; they were then pale greyish-ochreous on the back, brownish in the middle, through which ran a fine whitish dorsal line enclosed by two black outlines, which were united thrice across the white line within, at the beginning of each segment, suggestive of an irregular chain pattern. The four tubercular spots black and distinct. The sides, belly, and legs darker, of dull olive-brownish, a fine dirty-whitish longitudinal line above the legs, the subdorsal line darker brown, but interrupted near the middle of each segment by a short oblique streak of pale ochreous from the back, a fine, rather obscure darker brown line running midway along the side. Head ochreous.

Early in September they had attained their full growth, being about one inch and five-eighths in length, rather thick and cylindrical, the head, second and anal segments but a trifle smaller. The head shining olive-brown ; on the second segment a blackish-grey plate, rounded behind, through which pass the pale dorsal and subdorsal lines. The ground-colour of the back and sides very pale ochreous ; a diamond-shape of dark grey mottled with olive on each segment, having on its edges the usual four tubercular dots, rather large, and whitish ringed with blackish-grey ; the dorsal line whitish edged with blackish-grey, but more or less suffused with dark grey and olive as it passes through the middle of the diamond-shapes, though distinct at the beginning of each segment. A very short blackish-grey linear mark on the anterior part of each segment on either side. Subdorsal line whitish, outlined with dark grey, and in its course sometimes touching each posterior pair of dorsal dots. The sides are mottled with dark grey and olive, having oblique streaks and a pale line of ochreous above the legs. Spiracles brownish-orange, edged with dark grey; belly and legs pale greenish-brown.

Some of the moths appeared late in autumn, and others in the following June. (W. B., July, 1866; E.M.M., June, 1867, IV, 14.)

HADENA ADUSTA.

Plate XCIII, fig. 1.

The larvæ of this species seem easy to rear as far as their full growth, on lettuce, knotgrass, hawthorn, and sallow; about the end of September they become torpid, and hibernate until the warmth of spring awakens them, when they spin a slight cocoon under moss and dead leaves, the perfect insects appearing towards the end of June; but it frequently happens that in confinement the larvæ die during hibernation, as it is difficult to prevent their being attacked with mildew if kept moist, and, on the other hand, they die off if too dry.

The full-grown larva is about an inch and a half to an inch and five-eighths in length, cylindrical, and of nearly uniform width, of a full green colour, with the whole upper surface of the back and sides as far as the spiracles freckled with a deep purplish-red, which down the middle of the back becomes aggregated in the form of diamonds, each occupying the area of a segment within the subdorsal lines; these last, and the spiracular line with the space between them, are freckled and streaked obliquely with the same red colour. The dorsal line is only indicated by a dusky spot at each segmental division; the tubercular dots are blackish. The spiracles are white, edged with black; the belly and legs of the green ground-colour.

I am indebted to the kindness of Mr. Steele, of Congleton, for the subjoined varieties:

Variety 1.—Ground-colour a brilliant yellow, the

upper surface as above mentioned, suffused with deep
rose-pink; the dorsal stripe composed of two darker
pink lines, confluent at the beginning of each segment,
forming a spot; the subdorsal stripe bright yellow,
only visible on the anterior halves of the segments;
the tubercular spots and two transverse streaks near
the end of each segment also of the bright yellow
ground-colour.

Variety 2.—A dull pale yellowish-green, the dorsal
stripe faintly outlined with orange-red, with a spot at
the segmental divisions; subdorsal line of same colour,
but interrupted on the hinder half of each segment;
tubercular dots red, and situated on the faint reddish
outlines of diamonds, which are very delicately
freckled within; spiracles as in the others, white
ringed with black. (W. B., 1867; E.M.M., August,
1867, IV, 62.)

HADENA PROTEA.

Plate XCIII, fig. 2.

On the 7th of November, 1882, I had a dozen eggs
and one of *H. protea* from Messrs J. and W. Davis, of
Dartford. They were laid on a piece of chip, side by
side in an irregular group.

The egg is of medium *Noctua* size, rounded above
and a little depressed beneath, and yellowish when first
laid, judging from an infertile egg amongst them. The
thirteen fertile eggs when I received them were of a deep
rich brown inclining to chocolate or russet-brown,
having eight stout ribs, long and of a much paler tint
of the same, almost meeting at the top and forming
a star-like pattern, and between each of them low down
on the side is a short rib of the same tint. During the
winter they became rather dingier, and on the 1st of
May of a drab colour, and on the 2nd a paler tint of
drab, and by noon one hatched, three more by next

morning, one on the 4th, four on the morning of the
5th, and one on the 7th.

The newly-hatched larva is of a dingy opaque
bluish-green, the head marked with dark brown on the
crown of each lobe, and somewhat of a plate of
brownish appears on the second segment. After
moulting the larva is an active and robust little fellow
of a deep dingy glossy green colour, with black
shining head and narrow plate on the second segment.
By the 22nd of May the larva had grown to be five
lines long, the body tapering a little in front and
rather more behind, being rather stout in the middle ;
the head black and glossy, also a narrow plate across
the middle of the second segment, a blackish-brown
plate on the anal flap, the outside of the anal legs of
the same dark colour, the anterior legs black ; the
ground-colour of the body is a shining deep ochreous-
green, with a faint yellow dorsal line, a fainter and
much finer yellowish subdorsal, and a similar spiracular
line ; the fringes of black hooks on the ventral feet are
plainly visible. On the 25th it moulted, and on the
27th was grown to nine lines in length and quite stout
in proportion ; the ground-colour of the body was
light yellowish-green, the dark plate on the second
segment having disappeared ; a primrose-yellow dorsal
narrow stripe runs thoughout to the end of the anal
flap ; there still can just be discerned the finest thread
of the same pale yellow subdorsal line and also a
spiracular line ; the head is now brownish-olive, and
also the anterior legs. On the 30th it had moulted
again and was thicker, the head lightish green and
glossy, the body very pale yellowish-green, the dorsal
line paler and distinct, the other *fine* lines can only
be discerned with a lens. By the 2nd of June it had
grown very much, and stretched itself when crawling
to an inch and five lines. (W. B. ; May, 1883 ; N.B.,
IV, 181.)

HADENA DENTINA.

Plate XCIII, fig. 4.

On the 5th of July, 1876, Mr. William R. Jeffrey, of Ashford, Kent, sent me a single egg laid on a spikelet of a blossom of *Aira flexuosa,* which he found while examining the blossom.   The egg is of the usual *Noctua* form, round, domed at the top, and flattened at the base on which it is laid, strongly ribbed and reticulated, of a light salmon colour, blotched on the summit and encircled with a zone near the top of purplish-pink, with the surface glistening.   On the 7th it began to change wholly of a pinkish-grey colour, and on the morning of the 8th it hatched.

The young larva is grey, or rather greenish-grey, with the head and second segment ochreous-yellow, the body having conspicuous blackish tubercular dots and hairs.   I placed it on some *Aira flexuosa,* but put with it a few leaves of *Medicago sativa,* and on this it certainly fed and continued to feed, and on the 15th had moulted and assumed a green coat, its tubercular blackish-brown dots conspicuous as before. By the 21st it had again moulted and was now pale whity-brown, a still paler dorsal line visible and the tubercular dots warty and brown.   On the 23rd its present dress became fully developed, when each of the tubercular dots was conspicuously surrounded with a paler ring or halo, and a pale almost whitish dorsal line thickened towards the end of each segment became visible, also a paler subdorsal line, rather less white than the dorsal, edged above with darker, and at a short distance followed by another such pale whitish line; then follows a broad stripe of purplish-brown, darker than the back, on the lower edge of which are the spiracles, black with a pale centre; beneath is the pale flesh-coloured sub-spiracular stripe; the belly and legs rather deeper in tint and more ochreous, the thoracic segments

paler than the rest, and rather greenish-yellowish. The
dorsal line passes through a pear-shaped rather deeper
brown mark on each segment. On the 29th the larva
lay up to moult. Food latterly lucerne and knot-
grass.

On the 30th of July it moulted a third time, and was
now given knotgrass, *Crepis*, and *Hieracium*. Its
colour now is a purplish-brown with whitish dorsal and
subdorsal lines, the latter margined above at the beginning of each segment with a black oblong dash, the
blackish-brown tubercular dots distinct, each with a
plainly visible bristle. I was still unable to identify
this larva and once more speculating as to the probability of its being some *Agrotis* I had not known.

The larva moulted again in the night of the 7th of
August, and was now easy to identify by my figures
of older larvæ, being in its fourth coat very different in
appearance. It was half an inch in length, stout in
proportion, tapering but little in front, and rather
more behind. In colour it was now very dingy; the
ground-colour of brownish-grey was on the back and
sides very much obscured by dull blackish-grey dorsal
broad diamond-shapes of close freckles, relieved by
warty ground-coloured tubercular spots, each with a
dark centre and bristle; the dorsal line of brownish-
grey ground-colour appears as a very interrupted line
of faint freckles, enclosed just at the beginning only
of a segment within two squarish black spots; the
subdorsal line is a thin line of grey freckles, margined
above just at the beginning of a segment with a short
thick oblong dash of velvety-black; the side is closely
freckled with dark blackish-grey like the back, relieved by a much-broken line of brownish ground-
coloured freckles along the middle and by two tubercular warty spots; spiracles black; the subspiracular
stripe is grey somewhat freckled with darker; the
belly rather paler than the back, and tinged with
pinkish, but freckled like the back and sides; the
head rather pinkish-grey, streaked broadly down each

lobe and coarsely reticulated with blackish; anterior
legs pinkish, the others pinkish-grey, rather shining,
as also is the head, but for the rest there is no gloss, it
having a dull blackish and almost rough appearance.
The larva now fed slowly and liked garden lettuce
when given to it, but fearing to scour it too much I
continued to give it plantain, knotgrass, etc., but it
ate very little indeed, and I saw on the 17th it had a
conglomeration at its tail. It had grown much
blacker lately, and seemed inclined to hibernate. On
the 20th it died, and then when too late I regretted
not having given it lettuce entirely. (W. B., August,
1876; N.B., III, 107 and 131.)

HADENA SUASA.

Plate XCIV, fig. 2.

I have been much indebted to the assiduity of Mr.
Batty, of Sheffield, who kindly sent me, in July last,
some larvæ of this species in different stages of growth,
reared from eggs on broad-leaved plantain; they,
however, seemed afterwards equally partial to *Polygonum aviculare*, and fed up rapidly, retaining their
colours and markings throughout their growth. They
were full-fed and had gone to earth by the 28th of July,
and on the 23rd of August one moth emerged, greatly
to my surprise, and is a dark smoky-brown specimen.

The larvæ are uniformly cylindrical until nearly
full-grown, and then become a little tapering towards
the head; ground-colour green or yellowish-green,
most minutely irrorated more or less with yellowish
atoms; the dorsal line rather indistinct, and slightly
darker than the ground-colour; the subdorsal line
absent in some, but present in others as a fine black,
rather oblique streak on each segment, terminating at
the hindermost tubercular dot; the dots are black,
and arranged on the back in the usual trapezoid form;
the spiracular line is composed of a black fusiform

mark on each segment, with the white spiracles on their lower edges, brilliantly contrasted by a stripe of bright yellow below, along the side; the belly and feet green; the head and dull plate on second segment brownish.

There is a variety in which the ground-colour is brown or olive-brown, the dorsal line strongly marked as a double dark brown line vanishing at the hind part of each segment in some, and in others running continuously through a darkish brown triangle or diamond on each segment; individuals occurring with either form.

The subdorsal appears as a series of oblique dark brown streaks, each streak commencing a little on one side of the segment in advance, and abruptly terminated on the next by the hinder trapezoidal tubercular dot; the dots are black, and placed on small yellowish specks. Along the side of each segment is a triangular shape of dark brown, their bases bounded by the black fusiform marks containing the white spiracles, and immediately followed by a bright yellow stripe, its lower edge gradually tinged with the brownish colour of the belly and legs. (W. B., August, 1866; E.M.M., November, 1866, III, 136.)

HADENA THALASSINA.

Plate XCIV, fig. 5.

Early in July, 1865, Mr. Doubleday kindly sent me several examples of the larva of this species that had been reared from eggs, and were fed with *Polygonum aviculare*; there were several varieties in colour, but not a green one, as quoted from Freyer in Stainton's 'Manual.'

They were full-fed early in August, and the moths appeared from May 31st to June 4th, 1866, the next year; they were remarkably fine specimens.

The larvæ were about an inch and a quarter to an

inch and a half in length, cylindrical, and uniformly plump, the head being rather smaller than the second segment. The following are the three varieties amongst them.

Variety 1.—Reddish-brown above as far as the spiracles; a dull brown plate on the second segment, through which the dorsal and subdorsal lines are traced; the dorsal line pale ochreous on the anterior segments, but on the others much suffused with the ground-colour, except at the segmental divisions, where it reappears as an ochreous spot. The subdorsal line ochreous, and much suffused with brown. On the fifth segment to the twelfth, inclusive, a dorsal diamond-shape of mottled brown, darker than the ground-colour, and on each side a wedge-shape of very dark brown pointing forwards, their broad ends a little distance from the segmental divisions, their sides edging the lower half of the diamonds and the sub-dorsal lines. The wedge marks gradually increase in size towards the twelfth segment, where they are largest and darkest, and most conspicuous, by the subdorsal line being there suddenly paler, and united by a transverse pale line at the base of the wedges. Spiracular line black, on which are the white spiracles, and running immediately beneath is a pale greyish stripe, its upper edge whitish, belly and legs brownish-grey, head pale brown.

Variety 2.—A rich cinnamon-brown, mottled with ochreous above; belly and legs paler and greenish-ochreous; dorsal and subdorsal lines paler than the ground-colour, but not very distinct, the diamond marks hardly visible; the blackish wedge marks strongly defined, but with the addition of two or three fine streaks of ground-colour cutting transversely through them all; the tubercular dots black, in the following order: a transverse row of eight dots on the third and fourth segments, and on the fifth to the twelfth inclusive, the anterior dorsal pair distinct, the posterior pair hardly visible by being placed in the

broad ends of the wedges, a lateral anterior dot midway between the subdorsal and black spiracular line;
a dull brown plate on the second segment; head
brownish-ochreous, with a blackish stripe on each lobe
from the crown to the mouth.

Variety 3.—A dull greyish-brown; the dorsal and
subdorsal lines, and penultimate transverse mark, very
little paler than the ground; the tubercular dots black;
the wedge marks black, with a thin transverse line of
ground-colour cutting them through towards the
broad end. (W. B., 1867; E.M.M., August, 1867,
IV, 63.)

HADENA GENISTÆ.

Plate XCV, fig. 2.

By the kindness of Mr. Doubleday, who liberally
supplied me with part of a young brood reared from
eggs, I have been able to take figures and notes of this
species.

They fed well on *Alsine media* and *Polygonum persicaria*, and were one-third grown by July 14th, and by
the 29th some had obtained their maximum size, and
the others by August 7th, and had all retired below
the earth by the 14th. The moths came forth from
June 11th to 13th in the following year, 1866.

When young the colours of these larvæ were
brighter and darker than they afterwards became, with
distinct paler dorsal and subdorsal lines outlined
with darker, and black spiracular lines; otherwise
their markings were similar to the following.

When full-grown they were very plump creatures,
varying from an inch and five-eighths to an inch and
three-quarters in length, cylindrical, and tapering towards the head; the back and sides, as far as the
row of spiracles, of very mottled dull brown, brownish-
grey, dull greenish-grey, deep purplish-brown, or
dirty olive-greenish, for all these tints were found in

the brood. The dorsal and subdorsal lines outlined
with darker brown, in many instances only visible
on the anterior segments, and in others also at the
segmental divisions.

A series of darker brown diamond- and wedge-
shaped marks down the middle of the back, on the fifth
to the twelfth segments inclusive, viz., on each of those
segments a diamond united to a wedge-shape on either
side, the broad ends of the wedges extending to the end
of the twelfth segment only, and to about one-third
from the ends of the other segments, each wedge pointing forwards and reaching a third into the segment
in advance. The tubercular dots blackish, the upper
pair placed on the edges of the diamonds, the lower
pair on the broad ends of the wedges ; in the purplish-
brown variety the dots and lines paler than the ground-
colour, and in some instances not visible.

The whitish spiracles edged with blackish are placed
along the terminal line of the above brown colouring,
and the remaining surface below, including the legs, is
of a dirty whitish or pale drab colour, the legs tipped
with brown.

Head with two central black streaks across the
face ; a dark brown plate on the second segment, sometimes marked with one pair and in others two pairs of
pale spots. (W. B., 1867; E.M.M., August, 1867,
IV, 61.)

HADENA RECTILINEA.

Plate XCV, fig. 3.

On the 23rd of September, 1864, Mr. Doubleday
most kindly presented me with several nearly full-
grown larvæ of this species; and subsequently Mr. N.
Cooke obliged me with another variety, reared from
the same batch of eggs, which he obtained from a
female taken at Rannoch last summer.

They continued to feed well as long as any sallow

(the plant upon which they were feeding when I received them) could be procured for them; and about the end of October one individual retired to a slight hollow on the surface of the earth, and spun itself over with a semi-transparent web, in which it is now (at the end of March, 1865) still coiled up and visible.

One or two of the others died in February; the rest are still attached to the top of their cage, and during the last few weeks have occasionally crawled about a little, but apparently without feeding on any of the various leaves and willow shoots supplied to them; they have now lost much of their sleekness, and the segmental divisions appear contracted. The following is a description of their condition at the end of September.

Larva slightly tapering towards the head, with a transparent ridge or lump across the back of the twelfth segment. The dorsal and subdorsal lines grey, and very thin, edged with blackish; and the inner edge of the subdorsal, at the front of each segment from the fourth, diverges, thus forming a dusky oblique streak on each side of the segment, pointing behind to the centre of the back on the segmental division. The colours of the back are of rich dark browns, others of chestnut browns, and others of ochreous and orange browns. On the back of each segment, from the fourth to the twelfth inclusive, are paler wedge-shaped streaks on each side, viz., a short one on each side anteriorly, pointing backwards, and a long one on each side posteriorly, pointing forwards, thus giving somewhat of a diamond form to the darkest brown of the back. Ordinary spots pale, each containing a central black dot, and placed more on the anterior portion of each segment than is usual with the *Noctuina* in general. A broad deep brown stripe along the sides, mottled and streaked, and slightly edged below with darker colour, and along this are placed the very small white spiracles, which are edged with blackish. Above the first is a rather broad lateral

stripe, commencing next the head with a pale ochreous
or cream colour, blending gradually at the fourth segment into a grey-brown, and reappearing, similarly of
the same tint, on the twelfth and thirteenth segments,
and sides of the anal prolegs.

Belly dusky-brown.   Head blackish-brown and iri-
descent.

The sides of these velvety-looking creatures are
clothed with an excessively fine and soft pubescence,
which more or less includes the ventral prolegs, and
assumes the most delicate tints of bright azure, according to the angle of reflected light in which they are
seen.   (W. B., March, 1865 ; E.M.M., June, 1865,
II, 20.)

CLOANTHA SOLIDAGINIS.

Plate XCV, fig. 5.

For the opportunity of making acquaintance with
this long-wanted larva I am indebted to the kindness
of Mr. George Norman, of Forres, who sent me two
eggs on the 25th of April last.   These hatched in a
day or two, and the newly-emerged larvæ were of a
dark slaty-green colour, with the head dark brown.   By
May 4th they had attained to a quarter of an inch in
length, the ground-colour being purple, with the
dorsal and subdorsal lines pale grey, the (rather broad)
spiracular lines white, and the skin shining.   By May
24th half an inch had been reached, and the ground-
colour had changed to a very dark chocolate-brown,
with broad bright yellow spiracular stripes, the
narrow dorsal line being then of a pale slaty-blue.
Several days after this one of the larvæ died ; the
remaining one, however, fed on satisfactorily, and by
June 4th its colour had again considerably changed,
the ground being dark purplish-brown (darker on the
subdorsal than the dorsal region, which had a faint
pink tinge), the head dark brown, smooth and shining,

the dorsal stripe dark brown, with a distinct pale bluish-grey central line, but no perceptible subdorsal lines, only a broad, clear, pale yellow stripe along the spiracular region, the spiracles and trapezoidal dots grey, and the ventral surface, legs, and prolegs of a uniform dark purplish-brown. By the middle of June it was full-fed, and the adult larva may be described as follows :

Length about an inch and a half, and of average bulk in proportion. Head globular, the same width as the second segment. Body cylindrical, and of nearly uniform width throughout, being attenuated very slightly towards the head. Skin smooth and soft.

The ground-colour dark olive-brown, strongly tinged with purple. Head smooth and shining, pale brown ; the front of each lobe dark sienna-brown. Dorsal line dull slaty-blue, edged with smoke-colour ; no perceptible subdorsal lines, but a broad, clear pale yellow stripe along the region of the spiracles, edged on the upper side with a very fine black line, on which the reddish-brown spiracles were placed. On the front of the second segment a conspicuous black mark, and a transverse black mark on the hinder part of the twelfth segment. Trapezoidal dots very distinct, pale yellow. Ventral surface purplish-brown, tinged in the centre with green, gradually becoming darker towards the pale spiracular band. Legs brown and shining.

The larva, both in the adult and earlier stages, is very beautiful; the single one reared went down on June 19th.

At first the larvæ fed on whitethorn ; but, on being supplied with bilberry, evidently preferred that plant, which is, in all probability, the natural pabulum of the species. (Geo. T. Porritt, 10th August, 1872; E.M.M., September, 1872, IX, 92.)

XYLINA RHIZOLITHA.

Plate XCVI, fig. 3.

On the 26th April, 1874, I had the pleasure to receive from Mr. J. E. Fletcher, of Worcester, a few eggs of this species, which were laid on the 21st and 22nd of the month; and the larvæ were hatched on the first two days of May.

At first, and for some time, they continued to feed on the green cuticle of the tender young leaves of oak; but, as they grew, began at length to eat little holes through them.

The egg is small for the size of the moth, and in shape is spherical, but a little flattened; it cannot strictly be called ribbed, but is covered with thirty-five to forty longitudinal rows of pits in such regular order that their sides form both shallow ribs and transverse reticulations; in the centre of the upper surface is a button-like round spot ornamented with a star of nine pairs of short raised lines. The colour at first was almost white, the tinge of yellow being very slight; on the third day this turned to dull pink, afterwards blotched and streaked with pinkish-brown, at last becoming wholly brown.

The young larva is whitish with a buff-coloured head, until after the first moult, when by aid of a lens opaque white dots and hairs could be discerned on it. When not quite three weeks old the larva is half an inch long, of a greenish-white colour, showing distinctly the white raised dots and hairs. In four weeks it is three-quarters of an inch long and stout in proportion, of a rather pale bluish-green colour finely freckled with whitish, and having slight indications of dorsal and subdorsal lines; by this time it feeds well, eating through the leaves from the edges.

The full-grown larva measures one inch and a quarter in length, or a trifle more when stretched out in walking. It is of uniform stoutness, and cylin-

drical in figure, the head full and rounded, the hinder
extremity also rounded, and but little tapered; all the
legs are moderately well developed, and terminated by
sharp hooks. The ground-colour is a rather transparent pale bluish-green, appearing colder on the back
and sides than it really is, from being thickly sprinkled
over with minute opaque whitish freckles; these,
however, are but sparingly seen on the belly, which is
of a rather yellower green; the head is of a more tender
green, with a patch of paler freckles on the side of
each lobe; on the back of the second segment are four
whitish dots; on the rest of the body the opaque
whitish dorsal line is finely edged with darker green
than the ground, but is so much interrupted as only to
appear just at either end of each segment; the subdorsal shows similarly as a broken whitish line, and
less conspicuous, while the spiracular line is indicated
still more faintly, existing as an interrupted series of
larger whitish freckles than those which besprinkle
the skin; the wart-like tubercular dots are opaque
whitish, each having round the base a narrow unfreckled ring of the semi-transparent green ground-
colour, and each bearing a fine whitish hair; the spiracles white, delicately outlined with black; the terminal hooks of the legs whity-brown.

By June 3rd they had attained their greatest
dimensions, and by the 7th had ceased to feed, and
were become irritable, some having lost all their white
markings and turned wholly green like the colour of the
oak leaves, and by the evening they had retired into
some light soil supplied to them, where they spun up
in cocoons, and the moths appeared from September
28th to October 7th.

I found the cocoons were about three inches below
the surface of the soil, and they were composed chiefly
of fibrous particles spun together, and smoothly lined
with pale grey silk. The pupa itself is nearly five-
eighths of an inch long, and stout in proportion,
being a quarter of an inch in diameter; the head and

thorax rounded, the wing-covers long, the tip of the abdomen rather bluntly rounded off, having at the end a small rough knob furnished with two small spikes curving a little outwards towards their extremities ; it is of a mahogany-brown colour and very glossy. (W. B., 30th September, 1875 ; E.M.M., November, 1875, XII, 140.)

XYLINA SEMIBRUNNEA.

Plate XCVI, fig. 4.

On the 3rd of May, 1870, Mrs. Hutchinson kindly sent me sixteen eggs laid by a female moth on ash-twigs.

The egg is hemispherical, flattened beneath, rounded above, most minutely ribbed and reticulated, and of a cream colour. By the 12th of the month this colour began to change to grey. On the 17th a further supply of eggs was sent me by this lady ; they began to hatch on the 24th, and in the course of a few more days most of the young larvæ were out of the shells and feeding on the young ash-leaves, soon becoming a quarter of an inch long, at which time they were of a pale yellowish watery green, very pellucid, but with a faint opaque yellowish-white dorsal line just visible ; after moulting the subdorsal began in a few days to appear as a fine line, and by the 5th of June these lines had become more distinct. By the 12th the larvæ were five-eighths of an inch in length, and the tubercular dots were then distinct. A few days more and the most forward individual was an inch long. Their progress was very satisfactory, for by the 20th of June they had all attained their full growth, and by the 28th had retired to earth. The moths appeared from the 21st to the 27th of the September following.

The full-grown larva measured an inch and a quarter to an inch and three-eighths in length, and was cylin-

drical, but tapering slightly from the fourth segment
to the head, which is full and rounded. The last
three segments taper also a little to the anal tip. The
legs are well-developed, and the ventral and anal ones
furnished with sharp hooks. It is of a delicate pale
and bright yellow-green, the head rather pale bluish-
green, faintly reticulated with darker; the very distinct rather broad dorsal stripe of pale whitish-yellow
is a little attenuated or thinner towards the head and
on the last segment; the subdorsal line is well defined
only on the second segment, as is also the beginning
of another faint line below it; those on the rest of
the body are very thin and faint, composed of little
short whitish streaks in a line with each other but
much interrupted, the rest of the green ground-colour
of the back, except the second segment and the anal
flap, being very finely freckled with yellowish atoms.
The tubercular dots are whitish-yellow, set in a ring
of unfreckled ground-colour or green; the green
deepens a little into a softened line along the spiracles,
which are oval and white, delicately outlined with
black; the spiracular broadish stripe is pale sulphur-
yellow, and extends down the front of the anal legs;
the legs and ventral surface are similar to the back in
colour.

A day or two before retiring to earth the larva
becomes suffused with brown, deepening at the last
to a purplish-brown. (W. B., September, 1870;
N.B., II, 180.)

XYLINA CONFORMIS (= FURCIFERA).

Plate XCVI, fig. 6.

I have lately had the great gratification of rearing this
rare British species from the egg, and have figured the
larva at various periods of its growth. The eggs were
obtained from moths captured in Wales by a kind

friend (Mr. Evan John, of Llantrissant), who generously shared his good luck with myself and others.

Six moths were captured in October, 1870, and
were kept together in confinement through the winter;
and towards the end of February and the beginning
of March, 1871, eggs were laid by one of the females,
but the time of pairing was not observed.

The larvæ began to hatch on April 17th, the last of
them appearing on the 30th. They fed on alder
(*Alnus glutinosa*), and those that lived so long were
full-grown from the 11th to the 17th of June; but a
great many died off after their last moult, and I
fancied that, in the case of the larvæ which I fed myself, this mishap was caused by the alder leaves being
smothered with the secretion of the aphides, which
thickly swarmed on them. The pupa state lasted till
August; the first moth of which I have any record appearing on the 7th of that month, and the last on the
17th.

The egg is small for the size of the moth, globular
in shape, the shell thin, with about thirty fine ribs, and
irregularly reticulated between them; the colour at
first a pale straw-yellow, afterwards a dingy pinkish,
and lastly a dull purplish-brown, assimilating well with
the rough specks on the alder bark.

The larva escapes by an irregular hole in the side of
the egg, and at first is of a pale drab tint, and semi-
translucent, with the alimentary canal showing as an
internal green stripe. At first, and for three weeks of its
life, it lives and feeds within the hollows between the
ribs of the partially-expanded young alder leaves; by
degrees, as it feeds and grows, becoming more opaque,
and greenish in tint. When about a fortnight old the
colour is pellucid green, and distinct whitish longitudinal lines appear. In another week the colour is a
full bright green, and the lines whitish-yellow. At
the end of the month the length attained is fully half
an inch; the colouring now is at its brightest, the
ground being a rich velvety full green, and the lines

and tubercular dots bright sulphur-yellow. After this the growth is more rapid, and the colours become paler; when about three-quarters of an inch long, the colour is olive-brown, and the lines and dots pale yellow, namely, a dorsal stripe of uniform width, a subdorsal stripe rather broader, a fine wavy line between this and a narrow subspiracular line; the tubercular dots arranged in threes on either side the dorsal stripe. At the end of about six or seven weeks the final moult occurs, when the larva is about an inch in length, and with this moult the ground-colour becomes olive-green, and there come some black markings, giving an effect very different from that of the former stages; and I may observe that it was just at this time that the great mortality occurred, the larvæ, which hitherto had seemed to be doing well, now dying off one after another.

When full-grown, the length is an inch and a half, the figure rather stout in proportion, and cylindrical, except that the head is a trifle narrower than the second segment, which, with the third, also tapers slightly forwards, and that the thirteenth is tapered to the end; the head is full and rounded at the sides; the tubercular dots furnished with very small, fine hairs; the skin smooth and velvety. The ground-colour is olive-brown with a slight trace of green in it, particularly on the back; the sides and belly rather paler, having somewhat of a pinkish tinge. The pale yellow dorsal stripe is interrupted by a deep, blackish, freckled patch of the ground-colour, just at the beginning of each segment, which, by its extension backwards on either side, forms the dark boundary of more than half of a blunt diamond-shape of blackish freckles, the area within showing the yellow dorsal stripe but faintly; this dark freckling, with a deeper suffusion of ground-colour, forms a bar across the back from the hinder tubercular yellow dot on one side to that on the other; the part behind remaining to complete this irregular diamond-shape is but faintly freckled, and

there, at the end of the segment, the pale yellow
dorsal stripe shows bright and unclouded; on all the
segments, from the hinder tubercular dot, runs a thick
black streak, a little downwards and forwards into
the subdorsal pale yellow stripe, which it extinguishes
at that part nearly up to the segmental division, or
in some instances opens a little at one or at each
end, so as to allow the yellow stripe to appear. The
side, for about halfway or more down, is rather paler
than the back, then comes a very fine, rather wavy,
yellowish line, broken a little in character by black
atoms that make its edges appear ragged; the thin
subspiracular line is similar at a little distance below,
the interval being a little deeper in colour than the
side, and much freckled with deeper olive-brown; the
belly and legs are rather paler and a little tinged with
olive-pinkish, and bear some few freckles of yellow
and olive, sprinkled just above the ventral legs; these
last are tipped with pinkish-brown; the tubercular
dots are all pale yellow, and distinct, and are delicately ringed with black, as are also the oval, dirty-
whitish spiracles; the head is olive-brown, freckled
and reticulated with darker brown; the slightly more
shining second segment is, on the back, adorned with
two pairs of yellow dots.

When the larva ceases to feed, its habit is to retire
into moss, or, if it does not find this, it will fold up a
leaf, or else fasten a leaf loosely to the surface of the
soil, and there spin an oval cocoon, three-quarters of
an inch long, of whitish silk, close, but semi-transparent, and closely adhering to the surrounding sub-
stances.

The pupa has no striking peculiarity, being thick in
proportion, a little over five-eighths of an inch long;
the thorax, wing-, leg-, and antennæ-cases finely corru-
gated, and the abdominal segments rather smooth,
terminating in a hooked point, by which it is firmly
attached to one end of the cocoon; its colour dark
brown, the incisions of the segments brownish-red,

and the whole surface shining. (W. B., 11th September, 1871; E.M.M., VIII, 114, October, 1871.)

CUCULLIA VERBASCI.

Plate XCVII, fig. 1.

The larva of *C. verbasci* is similar in form to that of *C. scrophulariæ*, but rather larger and thicker when full-grown. The segmental divisions and wrinkles are marked with black interrupted streaks; the ground-colour is whitish, greenish-white, or bluish-green; a transverse, equally broad band of yellow, extending to below the spiracles on either side, is seen on the middle of each segment. This character is alone sufficient for its identity; and although this species varies much in colour and size of markings, yet the design remains in all.

In rudimentary marked varieties, the transverse central yellow band is often interrupted slightly on the centre of the back, and completely, or partially so, at the sides; the upper pair of dorsal black spots entire, and never united to those below. In richly marked individuals, the hinder pair of spots becomes elongated and they approach each other, with tails slightly turning upwards; in others not so confluent, a small twin pair of dots is seen instead on the yellow band in the centre, midway between the large spots. The yellow transverse bands are largely developed on the thoracic segments; the ordinary spots, dots, and streaks of black on the sides well developed. In some instances the ventral divisions are broadly black, and occasionally the whole surface of the belly is black. Perhaps hardly two larvæ could be found exactly alike in the minutiæ; but the transverse band of yellow is the conclusive character, strengthened by the additional one of the black anterior dorsal spots never being united to the posterior pair. (W. B., 1867; E.M.M., October, 1867, IV, 117.)

CUCULLIA SCROPHULARIÆ.

Plate XCVII, fig. 2.

In the following notes I hope to be of some service to those who, like myself, have entertained doubts concerning the real distinctness of this species and *Cucullia verbasci*, from inability to distinguish the larvæ found feeding on *Scrophularia aquatica* and *S. nodosa* from others on *Verbascum thapsus* and *V. nigrum*. It is therefore with great pleasure that I acknowledge my indebtedness to Mr. Doubleday, by whose kindness I am at length made acquainted with the real *C. scrophulariæ*, in four fine larvæ he presented me with on the 4th and 8th of last July (1867), feeding on flowers and seed-vessels of *Scrophularia nodosa*, the sight of which immediately dispelled all my previous doubts, as it did also any existing in the minds of Mr. Hellins and Mr. D'Orville, through whose hands they passed to mine; the latter gentleman having for years had great experience in, and devoted much attention to, the species of this particular genus in their larval state.

The larva of *C. scrophulariæ*, when full-grown, is an inch and five-eighths in length, plump, and cylindrical; the head rounded, and a trifle smaller than the second segment. Viewed sideways, it appears of uniform thickness; but seen on the back, it tapers behind from the tenth to the anal segment.

In looking on the back, its most valuable character, by which it can be instantly identified, is apparent in the bright yellow dorsal mark; for whether little or much intersected by black, it is distinctly seen to be a blunt-pointed triangle of yellow, close to the beginning of each segment, pointing forwards, its transverse base being longer than the sides, placed on rather less than the first half of each segment. The ground-colour in front of the two sides of the triangle, with belly and prolegs, is whitish-grey, or pale bluish-grey, or greenish-white; but the broad space behind the base of the triangle is a bright full green, varying individually towards bluish-green or grass-green. Thus it will be seen that there is a broad green band across the end of each segment. The black marks on the back may be regarded primarily as particular developments of the usual four spots, varying in each individual, and more or less like thick oval spots run together in blotchy marks; that is to say, each anterior spot is confluent only with the posterior one below it, but does not unite transversely with the others. In one variety the black spots resemble tadpole forms united by the tails; in another these tails are thickened equal to the spots, and appear as blotchy curves; and in one variety these blotchy curves are so thick and confluent as to include some of the ordinary side spots, thus completely surrounding two sides of the triangle with a blotchy black border.

To conclude the description briefly, there is a yellow spot on the spiracular region of each segment excepting the second; the usual black spots laterally and on the prolegs; occasionally some fine, short, transverse black streaks on the sides. The head bright ochreous yellow, mottled with red, and spotted with black; anterior legs reddish-yellow. (W. B., 1867; E.M.M., IV, 116, October, 1867.)

CUCULLIA GNAPHALII.

Plate XCVIII, fig. 2.

One larva from Mrs. Tester, found in Tilgate Forest, arrived on the 26th of August, 1871. It was feeding on the small lanceolate leaves of *Solidago virgaurea*, the leaves just below the flowers. It does not appear to eat the flowers. I figured it the same day. It was an inch and three-eighths in length, of moderate stoutness, cylindrical, though tapering a

little in the last three segments to the extremity,
the head rounded and scarcely less than the second
segment.   The skin of this larva is entirely without
gloss.   Its ground-colour is a rather deep green,
inclining a little to olive-green, and thickly freckled
with pale atoms of yellowish ; down the middle of the
back, from immediately behind the head to the anal
extremity, runs a dark purplish-brown stripe or band,
widest along the middle segments, bounded by the
subdorsal region ; this appears like colour or a stain
over the green ground, for it and the pale freckles are
faintly visible through it ; down the centre of this is
a series of dorsal diamond and roundish oval shapes,
two on each segment, the largest diamond or urn-
shape in front, the oval one joining it behind, faintly
deeper in tint, and finely outlined with black ; an
equally fine undulating black line runs on either side
of them in such a manner as to form by its enclosure
another series of diamonds along the segmental divisions, a trifle paler than the central or dorsal ones.
The transversely oval tubercular warts stand on the
boundary just within the dark area of purplish-brown.
The subdorsal line is merely the freckled ground-
colour defined by a freckled faint line of purplish-
brown, and followed by two others, very faint, sinuous,
and interrupted so as to be but little noticeable.   The
spiracles are of the green ground-colour, outlined
with black, each situated in a purplish-brown blotch
or slash placed obliquely along the side, and pointing
upwards and forwards.   Along the base of these runs
a longitudinal thread of interrupted pale yellow freckles,
becoming whitish as it approaches the head, on which
it continues to the side of the mouth.   The anterior
legs, as well as the ventral and anal ones, are green.
The ventral legs have each a spot of purplish-brown
on the side.   The back of the anal legs is of the
same colour, and their basal hooks are brown.   The
head is green, with groups of very minute black dots,
the ocelli black.   The tubercular fine hairs are brown ;

on the head and thoracic segments they point forwards, and on all the others backwards. The ventral surface is similar in colour to that of the sides, and freckled in the same way with pale yellowish, in a series of six longitudinal lines, rather thick and whitish in the central ones; that is, there is a broad central one and one equally broad on each side of it, and two other slender lines on each side of these, so that there are seven in all. The larva has a rough mealy look, which assimilates well to the rough stem of the plant. As it matures, the dark stripe on the back becomes paler, as though the green and pale yellow freckles showed more plainly through it; while the transverse subdividing wrinkles towards the segmental divisions become a little deeper and yellow when the larva bends its body round sideways.

By the 1st of September it had attained an inch and a half in length, and the dark colouring on the back had become a little paler. (W. B., September, 1871; N.B., I, 123.)

CUCULLIA UMBRATICA.

Plate XCVIII, fig. 5.

To the kindness of the Rev. Hugh A. Stowell and Mr. Greening I am greatly indebted for examples of the larvæ of this species, and for interesting details of their early history. The first-named gentleman captured a female at honeysuckle, which laid a large number of eggs on the 11th of July, 1866, and in five days they were hatched, and fed well on sow-thistles (*Sonchus*).

Unlike the sun-loving habits of others of the genus, these larvæ evinced a great aversion to light, and always hid themselves by day, reposing under the lower leaves of the sow-thistles, and at night ascending and feasting on the upper leaves and flowers

Those reared from eggs were full-fed by the 25th

of August, and the others by the 3rd of September,
and were kept separately and well supplied with
earth ; but instead of making subterranean cocoons,
they spun silken threads amongst the flower-buds of
the sow-thistles, attaching them to the tops of their
cages, and spinning under the buds a few threads,
forming a loose and open kind of hammock, in which
they changed to pupæ.

One individual chose a leaf curved downwards and
secured to the stem beneath by a few threads, amongst
which it underwent its transformation.

The pupæ were smooth and reddish-brown, with
the tips of the wing-cases projecting a little, and the
anal point considerably.

The larvæ, when viewed from above, tapered but
very little anteriorly or posteriorly, excepting the last
segment only, which was rather elongated, and depressed at an obtuse angle with the other segments.
The chief variation, individually, consisted of the more
or less suffusion of black, and of the degree of dul-
ness or brilliancy of the ground-colour. Amongst
them three examples will amply suffice for description,
the others being intermediate and connecting.

Variety 1.—Ground-colour bright ochreous-yellow,
with an elaborate blackish-brown raised and granu-
lated arabesque pattern of curves and angles on the
back ; the sides equally intricate, but linear and wavy
in character. The dorsal stripe is represented by
bare double triangular spaces of the ground-colour at
the segmental divisions, and on the last segment as a
central stripe. The subdorsal is indicated by a very
thin undulating line of the ground-colour, and on the
anal segment abruptly widening into a very broad
stripe, tapering to a point at the extremity. The
head dull black ; a dull blackish-brown plate on the
second segment, with three small spots of the ground-
colour on its front edge. Tubercular dots and spiracles
black, also the anterior legs and prolegs ; the latter
with a ring of white above their extremities.

Variety 2.—Ground-colour brilliant orange-ochreous, visible in spots at the segmental divisions along the centre of the back, and in narrow streaks along the subdorsal region, a much-interrupted line along the spiracles, and a row of spots and blotches on the side just above the legs (the larger blotches being above the anterior legs), and three broad stripes meeting at the end of the anal flap; all the rest blackish.

Variety 3.—Ground-colour dull brownish-ochreous, seen as dorsal, subdorsal, and lateral stripes, on the third and fourth segments with little interruptions, and on other segments only the faintest traces of them, excepting the anal, which is marked similarly to those previously described, and the dorsal stripe merely as a triangular spot at the end of the intermediate segments; all the rest of the body dull brownish-black, and each spiracle placed in a swelling blotch of intense and rather shining black. (W. B., 1867; E.M.M., February, 1867, III, 208.)

HELIOTHIS ARMIGERA.

The eggs of this species are extremely small for the size of the insect, nearly round and slightly striated, of a pale yellowish-green, becoming a trifle darker before hatching, which takes place in five or six days. As the parent moth continues to deposit a few eggs each night for a period of fourteen days, and probably for a longer time when at liberty, those first deposited are hatched, and change skins once or twice before the last eggs are laid. Some of the first larvæ feed up rapidly, and become imagos the same season; but the bulk lie over in pupæ till the following year.

The young larvæ are very sluggish, moving little, and eat only the lower surface of the leaf of the garden geranium or other food-plant. For the first fortnight they content themselves with this mode of feeding; they then commence to eat holes quite through the leaves, and no sooner is the hole suffi-ciently large to admit the head than they slowly crawl through it, only to eat another, and again and again repeat the process, so that they soon make a plant look as if it had been riddled with shot. They also now commence to eat round holes into the succu-lent shoots and stems, burrowing quite into the plant, and evince a strong liking for the buds and flowers. They would soon prove most unwelcome guests to any lover of his bright-flowered geranium beds. An entomologist would most likely be glad to sacrifice Flora to his aurelian pet; but a gardener would wage a war of extermination.

When about half grown the larvæ become terrible cannibals, eating their brothers or sisters with a zest and pertinacity quite horrible. They are mean and cowardly, generally seizing their weaker and more helpless brethren when about to cast their skins. As they became full-fed they appeared to hold each other in mortal fear, and, like most guilty people, lived in constant dread of being arrested for past offences, for when touched by another larva, ever so slightly, they would wriggle, twist, and throw themselves off the plant to escape a fate they had possibly inflicted on others.

When full-grown and extended they are about an inch and a half long, of moderate thickness, slightly attenuated from the middle, both anteriorly and posteriorly; the head is about the size of the anterior segment, shining brown, slightly mottled with darker shades; on the second segment is a coriaceous shiny plate or skin, giving it the appearance of being wet; the dorsal and medio-dorsal area is of a raw-sienna colour tinged with green, and pencilled in fine broken parallel lines of yellow and darker shades, varying a little in tone in different individuals, but

to no very great extent; there is a slight and interrupted dorsal line, formed by two fine oblong dark
spots, edged with yellow on each segment, and a still
more indistinct medio-dorsal line produced by four
or six dark-coloured small warts, two or three on
either side of each segment, and each emitting a short
bristly hair; the spiracular line is sharply defined, of a
pale ochreous, lined above, first with a fine yellow
and then a dark umber line, and below by a white
line; the legs and claspers are pale ochreous; ventral
surface a colourless grey, with three white lines.

The pupa is subterranean; and the moth appears
in August, September, and October. (W. H. Tug-
well, October, 1877; Entomologist, November, 1877,
X, 283.)

HELIOTHIS DIPSACEA.

Plate XCIX, fig. 3.

Greatly indebted for the help received from several
good entomologists, I here return my thanks to them
for all the opportunities they have so kindly afforded
me for studying the larvæ of this species, and, indeed,
without repeated help, I should have chronicled
nothing but failure; what with cannibalism amongst
the larvæ themselves, ichneumons, and drying up of
pupæ, out of eleven examples received at various
times I have reared but one moth, although I believe I have still some pupæ of 1873 alive.

My first acquaintance with the larvæ was in
August, 1867, when one was found in Gloucester-
shire, feeding on a blossom of purple clover, and sent
me by the Rev. E. Hallett Todd; I then guessed it
to be a *Heliothis* by its spiracles and texture of skin,
but, as it eventually died, its portrait remained among
the unknown, for future identification.

On the 25th of August, 1870, Mr. Harwood sent
me a similar larva, found in Norfolk, eating the seed-

capsules of *Silene otites ;* and on September 14th
another arrived from Lord Walsingham, with a notifi-
cation from him that he believed it to be *H. dipsacea ;*
this last was fed on sorrel for a few days, but did not
thrive, until some green seed-pods of toadflax were substituted, when a surprising improvement appeared in
its condition, and it soon grew to maturity ; but both
this and the other example died after spinning up for
pupation.

In August, 1873, my hopes were raised high by
the acquisition of several larvæ, found, and sent me
from Essex, by Mr. Harwood ; most of them he had
taken on *Ononis arvensis,* and they were nearly full-
fed, and soon retired into the sandy soil provided for
them, and there some of them still remain. The
last example I received, the one which has—by appearing in the perfect state—enabled me to identify all my
previous figures, was found on *Crepis virens* in Norfolk, and forwarded to me September 10th, 1873, by
Mr. W. H. Cole ; from this the moth appeared on the
10th of July, 1874.

From observing the habits of all these examples, I
conclude that the natural food of the larva, from near
half-growth onwards to maturity, is confined chiefly
to *flowers* and *unripe seeds* of various species of *Silene,
Ononis, Trifolium, Crepis, Hieracium, Linaria,* etc.

The full-grown larva when at rest is about an inch
and an eighth in length, and an inch and a quarter
when stretched out ; of moderate stoutness, the body,
thickest at the middle segments, tapers very little
towards the head, and rather more towards the anal
extremity, with a sudden slope down on the back
from the middle of the twelfth segment, the thirteenth
being rather elongated, and the anal legs extended
behind it, the other segments plump and well-defined ;
the head, which has rounded lobes, can be partly
withdrawn into the second segment ; the tubercular
dots small, each bearing a fine hair, and the skin is
partially roughened, as hereafter described.

The ground-colour is varied, straw-colour, light drab, greenish-ochreous, full green, brilliant yellowish-green, rather glaucous-green, olive-green, rose-pink, and deep purplish-brown having all occurred ; but in each individual the design has been the same in details as follows :—The head, often green but sometimes pinkish, is freckled with black or brown on the crown of each lobe; the dorsal line is the finest thread of ground-colour enclosed by a pair of much darker lines, which commence on the third segment, and thicken gradually as they approach the middle of the body, from whence they by degrees narrow again towards the end of it ; on either side of the back run two pairs of longitudinal, rather meandering, lines, a little darker than the ground-colour ; the sub-dorsal stripe of uniform width is either white throughout, or white on the second segment and afterwards pale yellow, or becoming faintly tinged with ground-colour, or else greenish throughout; when viewed sideways, it is seen to rise upwards a little in its course along the twelfth segment, and to form an angle by its sudden return to its former direction on the side of the anal flap, where it ends in a point ; immediately beneath this conspicuous stripe is a broad longitudinal band of ground-colour greatly filled up with darker colour than that of the back, its upper edge the darkest ; next below comes the spiracular line, either whitish, greenish, or pale yellow, and on it the *circular* white or pale ground-coloured spiracles, outlined with black, are placed ; then comes a stripe of ground-colour, or else ochreous or green, followed by a line of white, which runs down the front of the anal leg ; the belly is of the ground-colour, with a darker rather interrupted band above the legs, which are of the ground-colour, or else greenish. The texture of the skin in the darker lines and parts is rough, being composed of extremely short and minute bristly blackish points ; while in the intervals, and on all the pale stripes, it is smooth.

The pupa is five-eighths of an inch in length, of
moderate bulk, the head and palpi rather sharply produced, back of thorax swollen, wing-covers broad at
the ends ; abdomen tapering, and ending in two
longish anal points, the abdominal rings roughened
on the middle ; the colour a pinkish red-brown ; but
I see that the pupæ that are standing over to the
second year have become dark brown. The cocoon,
composed of silk of the weakest texture, is very flaccid,
but no doubt protects the pupa in the sandy soil.
(W. B., March 12th, 1875 ; E.M.M., April, 1875,
XI, 256.)

ANARTA MELANOPA.

Plate C, fig. 1.

For eggs of this, and of the following species also,
I am indebted to the kindness of Mr. J. T. Carrington, who sent them to me from Perthshire.

I received the eggs on June 4th, 1875 ; the larvæ
hatched on the 10th ; they soon began to feed on
tender leaves of *Arbutus unedo*, or *Luzula pilosa*,
sallow, flowers of *Helianthemum vulgare*, and on *Vacci-
nium vitis-idæa*, and by the 16th were growing and
thriving well. By July 3rd they were three-quarters
of an inch long, and feeding only on sallow, *Salix
capræa*, and *S. acuminata*, having gradually deserted
the other food-plants supplied to them; those that
now survived, some two or three only, continued to
feed till after the middle of the month, and about the
end of the third week in July turned to pupæ ; one of
them, without having attempted a cocoon, became a
bare pupa on the surface of the soil ; but as another
entered the earth, and apparently formed a cocoon,
we may suppose the latter would be the habit in a
state of nature.

The egg is almost globular, the shell delicate,
shining, with rather more than fifty ribs, the trans-

verse reticulation shallow, the top a little puckered; colour, when received, a delicate pink.

The newly-hatched larva has sixteen legs, but the ventral pair on the seventh segment are not serviceable, and those on the eighth smaller than those on the ninth and tenth; the usual warts small in size, and all placed on little eminences, and furnished with longish pale bristles; the colour semi-translucent whitish, but the back purplish, and the head pale brown, the warts black.

In about a week the legs on the eighth segment became nearly as much developed as those on the ninth and tenth, and those on the seventh increased in size; the whole body became greenish, the back brownish with pale central stripe, also a wider pale subdorsal stripe with a brownish thread through it. In about another fortnight the length attained was three-quarters of an inch, the figure of the usual *Noctua* type, tapering a little forwards from the fifth, and the thirteenth sloping rapidly; the skin soft and velvety; the ground-colour deep purplish-pink, dorsal line ochreous-brown boldly outlined with blackish, but interrupted on the fore-part of each segment by a reddish-brown triangular mark; this triangle is met on either side by a thick black wedge-shaped mark, below which again comes the continuous bright yellowish-white subdorsal line; this line is thin on the thoracic segments, but beyond them widens in such a manner that the widest part of it on each segment is near the end of the above-mentioned black wedges, and the whole line is finely edged with black throughout; the side is similar in colour to the back, but very much obscured by dark reddish-brown freckles, and with a short blackish streak slanting downwards on each segment; the spiracles oval and blackish; the subspiracular stripe yellowish white suffused beneath each spiracle with red, and delicately freckled with red along the middle; the belly and legs dark purplish-brown, the head also of this colour, with darker reticulations.

After the final moult the length became about seven-eighths of an inch, with the colouring much as before, except that the subdorsal line had become thinner, only just visible on the thoracic segments, and on the others much attenuated at each end, but still continuous. In about ten days from the final moult the full length was attained of somewhat over one inch and a quarter, the figure being slender for a *Noctua*; the ventral legs now all of one size; the bulk uniform; in general effect the appearance was less dark than before, though the details still remained the same, only the pale subspiracular stripe had become still more obscured by red and brown freckles; the tubercular dots of the back not noticeable, being situate within the black wedges; the spiracles now ochreous-brown finely outlined with black, and each placed on an unfreckled spot of the paler ground-colour; the belly mulberry colour; the whole surface velvety, except the head, which is hard and shining, and of a reddish-brown colour with darker reticulation, and a blackish streak down the front of each lobe.

The pupa, which lies exposed, is rather more than half an inch in length, smooth and rounded in figure, with the abdomen tapering off rather quickly, and ending in a blunt spike; very glossy, and in colour black, the segmental divisions being at first reddish. (J. Hellins, February, 1876; E.M.M., June, 1876, XIII, 11.)

ANARTA CORDIGERA.

Plate C, fig. 2.

The eggs which I received from Mr. J. T. Carrington on June 8th, 1875, were laid on June 1st, and the larvæ hatched on the 12th; meanwhile I had received from Mr. Buckler another supply of eggs, or rather newly-hatched larvæ, on the 10th, which had been sent him by Dr. F. Buchanan White. The young larvæ ate at first *Luzula pilosa*, *Arbutus unedo*, and *Arbutus uva-ursi*, the last kindly supplied by Dr. White; but after a time they were quite content with young leaves of *A. unedo*, and preferred them to those of *A. uva-ursi*, although I had been at the trouble of obtaining a fine growing plant from Messrs. Veitch, of Chelsea, in order to give them fresh tender leaves.

Both broods of larvæ grew and kept pace with those of *A. melanopa,* by July 3rd having become nearly three-quarters of an inch long, and by the 16th being full-fed, and retiring to earth about the 23rd.

The egg is about the size of that of *A. melanopa,* but not so globular; with about forty shallow ribs, and with faint transverse reticulations; the shell shining; the colour when laid cream-white, in a week becoming whitish with a faint reddish irregular ring and blotches.

The newly-hatched larva is of the same size as that of *A. melanopa,* but darker in colour, being pale dull purplish, with the head, collar, and anal plate shining blackish, the warts also blackish, distinct, and furnished with very short bristles, the ventral legs on the seventh and eighth segments small, and not usable. In about a week the colour changed to pale greenish, except the back, which was brownish, with pale dorsal and subdorsal stripes, the head and warts still remaining blackish.

At the end of the third week from hatching the larvæ were nearly three-quarters of an inch long, and all the ventral legs were used, those on the seventh and eighth segments, however, being still smaller than the others. The colour was now deep purplish-brown both above and below, with a white dorsal line and a faint indication of a subdorsal line, but only on the second and thirteenth; the subspiracular stripe pale primrose-yellow; the whole skin soft and velvety; the head horny. In another week, and after the final moult, the length was nearly an inch; the purple-

brown of the back now obscured by black, and on the
sides freckled both with black and with paler brown;
the dorsal whitish line thinner than before, and sometimes interrupted at the divisions by the ground-colour,
the subdorsal, though faint, now showing slightly all
its course; the subspiracular stripe becomes brownish-ochreous and freckled with crimson-brown, the
belly and legs dark purplish-brown; the head dark
purplish-brown, with a blackish blotch on the corner
of each lobe, hard and shining. At the end of the
fifth week from hatching the full length was attained
of one inch and three-sixteenths; the figure slender
for a *Noctua*; all the ventral legs about the same size;
in the colouring there were two varieties at least, and
perhaps in a larger number of examples more variation
might have been observed; the lighter variety had
the ground-colour crimson-brown, all the details much
as before, both the pale and the black freckles being
more distinct; the darker variety became almost black,
and had only a trace on the end of each segment of
the dorsal and subdorsal lines; the subspiracular
stripe was brown and tinged with deep lurid red; the
belly sooty-brown.

All the survivors of both broods, some four or five
in number, spun up in long rounded earthen cocoons
on the surface of the soil.

As a postscript to this and to the account of *A. melanopa,* I would say that from the information I have received from my friends, the natural food of *A. cordigera*
must be *Arbutus uva-ursi,* and that of *A. melanopa* probably *Menziesia cærulea,* but of this I am not sure; of
course *Arbutus unedo* and *Salix capræa* are only substitute foods. (John Hellins, February, 1876; E.M.M.,
June, 1876, XIII, 12.)

HELIODES ARBUTI.

Plate C, fig. 4.

It is with extreme gratification that I now find myself giving the history, from the egg, of this little sun-loving species, which I owe to the most kind and persevering help I had the pleasure to receive from Mr. H. T. Stainton in 1880, and again in 1881.

In the former year, on the 23rd of May, I received a cluster of about eight eggs, resulting from a moribund female after being a short time in a killing bottle of poison, but long enough, as it proved, to have destroyed their vitality.

On the 26th of the same month I was elated on receiving alive five captured examples of the moths; as two of them were females I imprisoned them and the most lively male together in a pot containing sprays of *Cerastium glomeratum* and *C. vulgatum* covered with leno, whereon they were occasionally fed with a drop of sugar and water, which the male imbibed plentifully, the females less often, and one of these soon left the leno and alighted on the *Cerastium*, and sat there with extended antennæ and wings gently vibrating as though intending to lay. The next day was dull and cloudy, and the two on the leno only flew around whenever a chance ray of sun gleamed on them, but late in the afternoon they made me hopeful of success when I saw they had paired about halfway down on the side of the pot, where they remained five hours and a half together; they were fed for five more days and fresh *Cerastium* added, but in vain, as they died without either female depositing even a single egg.

As a forlorn hope, I squeezed from the abdomen of the gravid and dead female several eggs; and after a few days I fancied one of them at least was changing colour, and in the afternoon of June 7th this one really began to hatch, and while noting down its details, which were well exposed to view, I could see the little larva making continual efforts to free its hindermost segment from a part of the shell adhering to the other eggs, but it was unable to extricate itself, and by next morning had perished.

With the return of May in 1881 I felt greatly encouraged to persevere, on finding that my previous failure had by no means diminished, but perhaps increased, the kindly interest taken by Mr. Stainton in the elucidation of the early stages of this insect, and he lost no time in giving me the result of his observations, both in literature and in the field; so that I soon learned what flowers were most visited by it,—for, as may well be supposed, some doubt of the food-plant had naturally by this time occurred to me,—and that *Cerastium arvense* was the plant assigned to *H. arbuti* by Carl von Tischer, who communicated this to Treitschke and afterwards to Freyer, as quoted by both, whom Guenée appears to have followed; I also learned that *C. arvense* does not grow in the district where *H. arbuti* is found flying by Mr. Stainton, but that *C. vulgatum* does, plentifully, of which he kindly sent me a few plants for potting on the 21st, and on the next day as many as twenty specimens of *H. arbuti*, all in lively condition.

The moths were distributed in three pots of growing plants, protected with glass cylinders and leno covers ; two of the pots contained the *C. vulgatum*, and the third pot some different plants of the Caryophylleæ, besides in each some tufts of *Bellis perennis*, whose blossoms constantly attracted and helped to nourish them, as did also sugar and water frequently supplied; in the evening of the 23rd I saw one egg had been laid on the glass cylinder, and on the 25th another egg on the opposite side of the same glass enclosing some of the *Cerastium*.

On the 1st of June a friend brought me some plants of *C. arvense* in full bloom, kindly obtained near Lewes, as the plant does not occur in this locality, and these were potted and protected with glass just in time for a second consignment of five living *H. arbuti* from Mr. Stainton, who yet in a day or two supplemented them with four more ; an egg was very soon laid on a leaf of *C. arvense*, and on the 7th I saw another egg was laid on the base of the calyx near the stalk of an expanded flower of one of the same plants ; these two eggs I cut off and sent to Mr. Hellins for his examination ; who had an accident which settled the first egg, and the second he pronounced to be addled.

Meantime I had often looked in one pot of *C. vulgatum* wherein no egg could ever be detected while the moths were alive nor after the cylinder was taken away ; yet on the 8th of June I was greatly delighted to see a larva quietly sitting on a stem, in an attitude rather suggestive of the letter S. After recovering equanimity from such an agreeable surprise, I became aware of a hole in the side of the seed capsule a little above it, and soon detected a second larva sitting quietly in the same manner, and then a third larva partly protruding from one of two contiguous capsules; and next, the hole in another capsule from whence the second larva had eaten its way out, like the first evidently soon to moult, a process they both accomplished in the evening of the 10th, and henceforward lived outside more or less exposed, feeding well on both flowers and unripe seeds; on the 13th I saw they were again waiting for another moult, which occurred a little before midnight of the 14th with one, and with the other at some early hour in the morn ensuing; they soon resumed feeding, and had grown decidedly by evening, and continued to eat quite voraciously, but less of flowers and more of seeds, eating out a number of capsules within a few hours, in this reminding me of the *Dianthæciæ*; they were full-fed by the 18th of June, when they left their food and lay up motionless for a day and night, as though to purge themselves of their grossness while secreting the needful silk before entering the earth for pupation.

These larvæ conveyed an instructive lesson in showing why I failed the year before to get any eggs laid on sprays of the food-plant when *gathered*, also on this occasion the wonderful instinct and prevision, I may say reasoning power, of the parent moth or moths, which refused to lay more than three eggs on the few plants confined with her or with them—for there remains the possibility that perhaps three females were confined, and each laid one egg, knowing there would be barely enough sustenance for a single larva. But, however this may have been, it would seem that in nature the female would deposit her eggs singly, probably in the corolla or on the calyx of a flower, just here and there one, in proportion to the abundance of the plant.

I know not if this larva had been seen by any human eye since the time of Carl von Tischer, but the time for it to be found in this country had come, for on the 17th of June I received a further very kind attention from Mr. Stainton in the arrival of a full-grown larva of *H. arbuti*, which he had gathered by chance while getting some *C. vulgatum* for a coleopteron in the field where *H. arbuti* flew; this larva in no way varied from those I had reared, and proved to be only twenty-four hours later in maturing. Curiously enough, this incident was repeated similarly by the Rev. John Hellins, to whom I had sent a larva of *H. arbuti* reared from an egg laid, I presume, *within* a flower of *C. arvense* (as after many repeated close searches I failed to find more than the two before mentioned on *C. arvense*), and he, returning home with some of that species for food on July 2nd, found a larva of *H. arbuti* emerging from one of the seed capsules he had gathered.

The egg of *Heliodes arbuti* is globular, about $\frac{2}{5}$ mm. in diameter, having a slight depression beneath; it seems thin-shelled and finely pitted all over, shining,

and is of full yellow colour, turning rather brownish
just before hatching on the seventh day.

The newly-hatched larva is white, with brown head
and a narrow brown plate on the second segment.
After living hidden within a seed-capsule and feeding
on the unripe contents for about from fifteen to seventeen days, during which it has got through its earliest
moultings and acquired a colouring that assimilates
most wonderfully well with that of the capsule of the
plant, as it waits outside for its penultimate moult ; it
has a brown head streaked and spotted with darker
brown, and the body is either of a pale watery-green
colour or slightly tinged with pinkish-grey, and
marked with a dark green dorsal line, a whitish subdorsal line, and a stouter white spiracular line, the
ventral legs clear and nearly colourless ; after this
moult it is nearly six millimetres long, the head and
second segment pale brown, with slightly darker brown
marks, the rest of the body much deeper and richer
coloured than before, either a greenish-grey or a pinkish-grey ground—as both varieties occur at this stage
—and now the dark slaty-green dorsal line runs in the
middle of a broad softened stripe of paler ground-
colour than the rest of the back and the side ; next
comes the whitish subdorsal line, and after an interval
of ground-colour the perfectly white spiracular stripe ;
both of these are very conspicuous. Though all the
ventral legs are equally well developed, it still often
assumes its former favourite position while resting,
which is very much like that of a half-looper, holding on
sometimes by the anal and the fourth pair of ventral legs
only, at other times with the addition of the third pair,
while the others and all the fore-part of the body are
held off free, with the head bending downwards, forming an arch. After feeding three days the ground-
colour is lighter and greener, and the length when
laid up is eleven millimetres.

After the last moult it attains in four days its full
growth, when the length is twenty millimetres and

stoutish in proportion, of true *Noctua* form, with
plump twelfth segment; the thoracic segments slightly
taper towards the smaller and rather flattened head;
the mouth prominent. In colour the head and plate
are of a light greenish tint and glossy, the ground of
the rest of the body is light green, the dorsal line dark
green, the whitish subdorsal line is finely edged above
with darker green than that of the back and side;
the yellowish or yellowish-white spiracular stripe is
well relieved along the upper margin by a conspicuous
dark green stripe; the spiracles are whitish, finely
outlined with black; the tubercular dots are brown,
but too minute for any but powerfully-assisted vision;
the belly and legs a rather paler green than the back,
the skin soft and smooth; when it has ceased to feed
and is laid up all the lines soon disappear, and it is then
of a uniform green colour.

The larva fabricates at about an inch or two beneath
the surface of the soil a cocoon of earth, with a thickness of wall about one millimetre, or in parts even
less, kneaded well together with silk, and slightly
attached to a few coarse particles of earth outside; it
is of close texture and not very brittle; the general
figure is roundish or roundish-oval, and it measures
about nine by six or seven millimetres; the interior
is very smooth, and just fits the pupa comfortably
without room to spare; the pupa itself is of a very
dumpy form, with rather a bluntly tapered abdomen,
having at the tip two fine thorny points of inconceiv-
able minuteness, and in contact with the compressed
old larval skin; in colour the pupa skin is reddish-
brown and rather shining, and in length six to seven
millimetres.

The perfect insects were bred, both male and female,
in the morning of the 4th, and a female on the 11th
of this month (May, 1882). (W. Buckler, 12th May,
1882; E.M.M., July, 1882, XIX, 36.)

AGROPHILA SULPHURALIS.

Plate C, fig. 5.

Hübner's figures of this species leave me little that is new to say about it ; still I feel much indebted to Mr. T. Brown, of Cambridge, for enabling me to rear a larva which Mr. Buckler has figured.

Unluckily, although the moth had laid several eggs, they all perished in the post-office save one, and the single larva did not live to become a pupa, having been hatched on June 25th, and dying on August 15th.

I potted for it a small plant of *Convolvulus arvensis*, and on two little shoots of this, bearing in all not more than five or six very small leaves, it fed and grew and moulted contentedly during the first half of its fifty days' life, its longest journey all that time not exceeding an inch and a half.

Had the other eggs escaped *squashing* on their journey, probably I might have had the pleasure of seeing both the varieties which Hübner figures, but the green one yet remains a desideratum ; my single larva was his brown variety.

When first hatched, it was a dingy-grey little looper, with a black transverse dorsal hump on each of the four middle segments, but at each moult these humps became less, till at last there remained nothing but the usual dorsal dots, black and distinct, and these too afterwards disappeared. When full-grown the larva is about an inch long ; the legs twelve ; the body cylindrical, thickest at the fourth segment ; the segmental divisions deeply indented ; when at rest the middle segments are generally arched, and the head bent down. The colour a rich chocolate-brown ; dorsal line rather darker, and edged with very fine paler lines ; subdorsal line also darker, but scarcely visible ; spiracular stripe broad, of a pale yellow, and with a fine brown thread running throughout its length ; immediately after the last moult there were some rich yellow and orange spots also in it, but these disappeared, and the whole stripe grew paler. (John Hellins, September 16th, 1867; E.M.M., October, 1867, IV, 115.)

ACONTIA LUCTUOSA.

Plate CI, fig. 1.

I am greatly indebted to Mr. Howard Vaughan for kindly giving me the opportunity of figuring and describing larvæ of this species, as well as for furnishing some interesting details concerning their earlier stages.

The eggs were laid on the 7th and 8th of June, 1868, and hatched on the 16th and 17th of the month.

The young larvæ at first appeared to be veritable loopers, twelve legs only being visible; but as they grew larger the other legs became apparent, though still in walking they did not use the first pair of ventral legs.

They appeared to be nocturnal feeders, eating the flowers and seeds, as well as the leaves, of *Convolvulus arvensis*; they reposed, lying along and closely embracing the stems of the food-plant, close to the ground, and in this position would easily escape observation.

The full-grown larva is about one inch and a quarter in length, slender, and stoutest in the middle, and tapering a little towards the head (which is smaller than the second segment), and more to the posterior extremity; the folds and divisions moderately indented on the first four or five segments, but hardly noticeable on the remainder. The two hinder pairs of ventral legs more developed than the two preceding pairs.

The ground-colour on the middle of the back is a *pale* greyish-ochreous, brownish-grey, or reddish-grey, the sides being darker and browner; the dorsal stripe tapers at each extremity of the larva, but is narrowest on the anterior segments, the stripe itself being of the pale ground-colour above mentioned, but faintly outlined interruptedly by short dots or lines of black; sometimes towards each segmental division it is delicately freckled with a slightly deeper tint of the same, and, in some examples, two short black streaks, rather thicker than those that outline the stripe, appear at the beginning of each segment, almost forming a v, pointing forwards.

The pale region of the back assumes a kind of chain pattern from being bounded on each side by a rather broad sinuous border of *dark* grey-brown, on which are placed the anterior pairs of tubercular dots, being large and very pale greyish, delicately margined with blackish; the posterior pairs small and black.

The subdorsal stripe is but little paler than the dark ground-colour of the sides, and chiefly towards the head, and just a little at the beginning of each segment, the stripe is edged with a line of dark brown; beneath this again come three other dark brown lines, the lowest of which is the spiracular, and is thicker than the others; the upper two are slightly sinuous, and the second bears a pale tubercular spot at the anterior part of each segment, and also touches the spiracular line in the middle of the segment.

The spiracles are black and circular. Below them is a broad stripe of very pale brownish-grey, edged above with a paler thread, and below with a little darker stripe of reddish or greyish brown, followed by another close above the legs of paler greyish-brown. The belly slightly deeper greyish-brown, with a central brown stripe bearing on the middle of each segment beyond the fourth a blackish round spot. Legs pale brownish-grey; pro-legs similar, and with a dark brown dot above their fringes.

The head slightly hairy, and very pale greyish, having on each side four lines of black dots in continuation of dark stripes on the body. The second segment has a semilunar dull dark brown plate, through which run conspicuously the dorsal and sub-dorsal pale stripes.

The pupa is subterranean. (William Buckler; E.M.M., August, 1868, V, 75.)

ERASTRIA FUSCULA.

Plate CI, fig. 3.

To Mr. G. C. Bignell, of Devonport, my best thanks are due, not only for kindly supplying me with the larva of this species in the autumn of 1873, but also for clearing up what had been the reason of my failing to procure it before.

One night in the autumn of 1857, the year in which I began collecting, I found a twelve-footed larva walking on the ground, which spun up at once, and, during the next summer, produced *E. fuscula.* Not having found it on its food, and seeing that the books with one consent gave bramble as the food, for many subsequent years I used to beat the brambles in the same locality, hoping to get more larvæ; and when I could take the moths I used to shut them up with bramble sprays in order to try for eggs. But in neither case were my efforts successful,—and why? In the autumn of 1873 Mr. Bignell, whilst sweeping herbage at night, took several larvæ off a stiff grass, *Molinia cærulea,* growing in damp places; these, on examination, he concluded to be *E. fuscula,* and the following summer proved his conclusion to be correct.

The secret of our previous puzzle is now out; one might have beaten brambles for ever without finding a larva.

The larvæ came to me on September 10th, 1873, and spun up by the end of the month; the moths appeared during the last week of May, 1874.

The full-grown larva is about three-quarters of an inch long, rather slender, and even in bulk throughout; the twelfth and thirteenth segments taper a little; the head full and round; fully-developed ventral legs on segments 9 and 10, with rudiments of legs on segment 8; in walking it is a semi-looper; the colour on the back is pale yellow with a broad greenish pulsating dorsal vessel; the subdorsal is a thin line of clear yellow edged above with brown, and below with greenish; the round black spiracles placed on a thin reddish line; anal legs sometimes purplish; the usual dots on the back blackish ringed with reddish; the belly yellow, with its dots black.

Some of the larvæ have a more reddish tint, and have every line edged with decided red; with a brownish stripe between the lower edging of the sub-dorsal and the spiracular line, and below this again a yellow line, then a red line, and the belly dull pale brownish.

The cocoon is very firmly and neatly made of a thin coating of silk, stuck all over with fine earth or sand, about four lines deep and two wide. Some spun among moss, by larvæ which died, were not so close or tough, and were both longer and wider.

The pupa is about five-sixteenths of an inch long, cylindrical, stoutish about the thorax, the abdomen smaller and short in proportion, ending rather bluntly in a spike set with several curled-topped spines; the pupa skin very glossy, rich red-brown; the wing-cases more golden-brown; the eyes blackish.

By the kind help of the Rev. T. A. Marshall I am able to add that the name of the ichneumon, which was bred about the middle of April from some of the cocoons, is *Protelus chrysophthalmus*. A saw-fly larva much resembling that of *E. fuscula* in colour feeds with it on the same grass, but I have not found out to what species it belongs; and I shall leave some one else to guess which of the two is the first wearer, and which the mimic, of the colours of their common dress. (J. Hellins, 14th July, 1874; E.M.M., August, 1874, XI, 66.)

BANKIA BANKIANA.

Plate CI, fig. 4.

This pretty and active little *Noctua*, of which nothing had been heard for a long interval of time, was in the season of 1882 re-discovered by Mr. (now the Rev.) G. H. Raynor,[*] who found it in some abundance near Ely, and succeeded in obtaining a good number of eggs, and most kindly sent a liberal supply of them to my friend the Rev. J. Hellins and myself; those I received were laid within a glass-topped box, to which they adhered, as well on the glass as on the paper, being sprinkled over both surfaces singly, with occasionally two together.

The eggs arrived on the 7th of June, 1882, and began to hatch on the 9th, while yet the exact nature of the proper food-plant for the larvæ seemed somewhat uncertain; a low plant had indeed been suggested to me by Mr. Raynor for trial, since although Guenée had distinctly stated *grasses* to be the food, he had not mentioned any particular species of grass; I soon found, however, the low plants refused, and then tried a small *Carex*; they fed a little on this and on coarse grasses, but the little larvæ began to die off; when, fortunately, before all had hatched out and died, it was found that *Poa annua*, a common grass growing almost everywhere, was quite to their taste, and the fact was at once kindly imparted to me both by Mr. Hellins and Mr. Raynor.

The larvæ throve very well on the *Poa* up to the third week in July, when, as often happens with this grass indoors, it was attacked by mould, which caused the death of almost all my larvæ; however, Mr. Hellins most kindly sent me several of his, which had been kept in the open air, so that I was able to con-

* Or rather by Messrs. W. Warren and Cross (see **E.M.M.**, October, 1883, p. 117).

tinue my observations until the end of the month, when the larvæ reached full growth.

I kept my pupæ alive through the winter, but suppose I mismanaged them during the month of May, 1883, by keeping them too much exposed to rain, as I bred only one specimen, a male, on the 29th of June.

The egg of *B. bankiana* is globular in shape, with a slight depression at the base, about one-thirty-fifth of an inch in width, and one-fiftieth of an inch in height, with about thirty-four shallow ribs, and with shallower transverse reticulations; the central space in the top is flat with large shallow reticulations; the shell has a pearly sheen; when first laid, it was said to be of a dull whitish, having the faintest greenish tinge, and then gradually turned to a pale greenish-yellow.

When first hatched the larva has the ventral legs developed on the ninth and tenth segments, and a small undeveloped pair on the eighth; it is of pale yellowish-green colour, with very fine black dots and hairs. After feeding a few hours the interior became deeply tinged with dark green, which showed strongly through the clear skin, especially in the middle of the body; when eight days old the skin became less clear, and of a uniform light yellowish-green with blackish tubercular dots.

In twelve or thirteen days they moulted the first time, and became less transparent than before; and after the second moult, in five or six days' time, they were long and slender, and of a more opaque velvety green, and faintly showed subdorsal lines of paler green.

After another week the third moult occurred, when the ground-colour was a little fresher than before, the head very pale green, and a dorsal line of darker green than the ground showed faintly here and there; the subdorsal lines were whitish-yellow, and also the segmental divisions, while the length had increased to seven and a half lines.

The fourth moult occurred on the 14th of July, and by the next day they had become nine lines long, and the small undeveloped pair of legs on the eighth segment were still to be noticed; the slender proportions of the larvæ, remarkable from the first, seemed now to be even more striking as they attained full growth towards the end of the month, when they measured from eleven to twelve lines in length; they were of a very yellow-green colour, with yellow segmental folds, the round head of a light green colour with upper lip whitish, and mouth black; the dorsal line dark green, though faint; the subdorsal stripe primrose-yellow; the roundish spiracles flesh-coloured, placed on the deep yellow thread-like trachea, showing faintly through the skin.

On the 1st of August one larva began to spin its cocoon just beneath the crown of the grass-roots, almost close to the surface of the earth; and others followed in the same way during the next four days, though one larva lingered two or three days longer; this was exactly an inch long as it lay stretched out, according to the habit of this species when at rest among the grass, which it matched in colour remarkably well.

The pupa is very short, stout, and dumpy, three and a half lines in length, the thorax and wing-covers well defined, the last rather long in proportion, and from them the abdomen tapers obtusely to the tip, which is furnished with two fine points and minute curly-topped bristles; its colour at first is of a light drab, but towards May of the year following it becomes a dark brownish-green, and is rather shining. (William Buckler, 24th July, 1883; E.M.M., September, 1883, XX, 77.)

HYDRELIA UNCANA.

Plate CI, fig. 5.

I am indebted to Mr. Carrington for eggs of this species. They were laid on June 23rd and 24th, 1868, and received by me on the 28th June.

The egg is soft-looking, rather irregularly shaped, but still of the usual echinus-like outline, with nearly forty very shallow and irregular ribs, connected by irregular transverse reticulations, and in colour a full yellow ; in fact, it looks like a little speck of butter.

On June 29th the eggs became dark grey, and on the 30th the larvæ came forth ; by the 17th of July they were about a third of an inch in length, by the 28th they were three-quarters of an inch, and by the third week of August full-grown.

They fed well on *Carex sylvatica* ; when at rest, stretched out flat along the blades of their food ; looping in walking, and jumping about angrily when touched.

The newly-hatched larva is a little greenish looper, with the usual dots showing brown, and emitting bristles. As it grows it becomes more and more of a full green after every moult. When it is full-grown the length is quite an inch, the figure slender, cylindrical, uniform throughout in bulk, save that the third segment seems a trifle swollen, and the last three segments taper slightly to the anal flap, which is bluntly rounded off, or almost squared off ; the head is hard and globular, about as wide as the second segment ; there are two pairs of ventral legs fully developed and usable, and the rudiments of another pair, useless.

The colour is a full velvety-green, with a pulsating dorsal vessel of a darker tint ; there is a fine whitish-green subdorsal line, and a rather broader spiracular line of very pale yellow ; the spiracles are indistinctly brownish, and the hinder segments paler than the rest of the back ; the belly is also paler, but still of a soft rich green ; the head somewhat yellowish-green.

The larvæ retired under ground for pupation. (John Hellins, December 14th, 1869 ; E.M.M., March, 1870, VI, 232.)

98 BREPHOS NOTHA.

BREPHOS NOTHA.

Plate CI, fig. 7.

I had no opportunity of becoming acquainted with
this species till 1869, when Mr. W. H. Harwood
kindly sent me several young larvæ; these fed well,
but as I did not know how to provide for their pupation, my hopes of seeing the imago in 1870 were sadly
blighted. However, in that year Mr. W. R. Jeffrey
sent me two larvæ from Saffron Walden, and as I
managed to accommodate them more suitably than
my former stock I succeeded in rearing two fine
moths.

As the insects appear early in April, the eggs must
be laid some time during that month; the larvæ feed
on aspen (*Populus tremula*), spinning the leaves
together flat-wise for concealment; those I had in
1869, on June 2nd, were still small, barely half an
inch in length, but they grew fast after this, and
retired to change by the 29th. The dates I have for
the appearance of the imago are April 8th and 9th,
1870 (both cripples), and April 4th and 7th, 1871.

The larva, up to half an inch in length, is very
dingy, nearly black, but bearing some exceedingly
fine, pale drab longitudinal lines; after moulting, and
when about three-quarters of an inch in length, it
becomes less like a *Noctua* in form than it was before,
and more like a *Geometer*, both in form and manner of
progression; its colour now is of a delicate green,
inclining in some instances to glaucous; the longitudinal lines become whitish-yellow; the head and second
segment spotted with black; the segmental folds
whitish-yellow. The growth now is rapid, and in
some individuals black spots appear on the sides, in a
day or two developing into stripes; but in others no
more spots appear than those on the head and second
segment.

The larva, when full-grown, is about one inch in

length, not very stout, cylindrical, and diminishes so very slightly towards the extremities, that it appears of uniform bulk throughout; the head is full and rounded; the two front pairs of ventral legs are much less developed than the next two pairs, and the hindmost pair are splayed laterally; beneath the anal flap is a small point, with a tubercle on each side of it; the segments are plump and well-defined; the mode of progression is an undulating, half-looping, quick walk, changed to what may be called a run when the larva is exposed to light.

The ground-colour is now of a pale subdued tint of green, or else a bright velvety yellowish-green, the dorsal vessel rather a deeper tint of the same, edged with fine lines of pale greenish-yellow; the subdorsal fine line is yellow, and between it and the spiracles runs another such fine line. The side being now more or less black is, by this pale line, divided into two broad black stripes, which in some specimens are complete, in others only partly so; the spiracles are white, outlined delicately with black, and beneath them is a broadish stripe of pale yellow or whitish-yellow; the belly and legs paler green than the back; in one individual a short, tapering, black streak issued at the end of each segment for half its length forwards as an edge to the subdorsal line, but these streaks began on the twelfth and ceased at the fifth segment. The black marks on the head and second segment appear to be constant, and characteristic of this species; they may be more minutely described as follows:—a broad irregular blotch down the front of each lobe, forked at the side, and a round spot on the face between them, and two pairs (sometimes more) of black dots on the second segment; the pale lines of the back are absent from the second segment and from the anal flap; the anterior legs generally dotted with black. One of Mr. Jeffrey's larvæ furnished a good variety; its colour was a dingy, rather olive-brown, with the lines of a pale pinkish-grey, with

only the usual black marks on the head, second segment, and anterior legs.

My first set of larvæ I furnished with earth and
moss for pupation, with the result of causing the
death of all but two, which produced crippled insects ;
the second set were supplied with pieces of bark, into
which they could bore, and I now understand that a
dead stick or piece of dry decaying wood would have
been better still, for the habit of the larva is to exca-
vate in the solid bark or wood a smooth cylindrical
chamber (reminding one much of the work of some of
the carpenter bees) just big enough—without the
least waste of space—to accommodate the pupa with
the shrivelled larva-skin behind ; the circular entrance
to the chamber is stopped with the gnawed raspings
of the wood mixed with silk, but there is no silken
lining to the chamber itself; the pupa lies with its
head towards the entrance, and, after the exit of the
moth, the empty pupa-skin remains in the chamber.

The pupa is about half an inch long, cylindrical,
and uniform, except a rapid tapering at the tail end,
and tolerably smooth except at the abdominal divisions, where there are rings of minute points ; the
abdomen terminates in a thick, blunt, somewhat flattened knob, furnished with two spikes, which, instead
of projecting as usual in the same line with the body
(or knob), turn off at right angles on either side ; the
colour a shining dark red-brown.   (W. B., June,
1872 ; E.M.M., July, 1872, IX, 41.)

PLUSIA ORICHALCEA.

In the month of July, 1882, in an outlying part of
the Cambridgeshire Fens, eight or ten worn specimens
of *Plusia orichalcea* were captured by the aid of a
lamp, hovering round flowers of *Eupatorium canna-
binum*; one only, the first specimen obtained, was
taken flying in the afternoon sunshine.   This year I

had the good fortune to beat ten specimens of a larva which, though exactly like that of *Pl. gamma*, but a little larger, produced in July nine beautiful *Pl. orichalcea.* Three others were obtained, one each by Messrs. Archer, Cross, and Raynor, of Ely, but were not reared. Of the earlier stages of the larvæ I cannot speak, as those beaten were all past, or near, their last moult. As far as I could see, their colour, size, and markings are exactly those of *Pl. gamma.* There are two fine white lines down the back from the third to the penultimate segment, with the dorsal vessel showing darker green between them; oblique white lateral lines on each of these segments. On the second and third segments and on the anal segment, there are five irregular white lines, which unite together in front, in the direction of the head. The spiracles are white and small, except the last, which is conspicuously larger than the rest. But the most striking feature of this larva is its wonderful power of extending and withdrawing the first three or four segments of its body, and reminding one of the larva of *Chærocampa elpenor*, or of the common earth-worm.

When full-fed the larvæ spun a flat oval pad of white silk on the side of the muslin bag in which they were reared, and thereon remained for twenty-four hours or longer, perfectly motionless, in a horseshoe-shaped form, the head in close proximity to the tail. After this interval of rest they proceeded to spin the rest of their cocoons, which were soon completed, being thin and transparent enough to allow of the easy observation of every movement of the larva inside. The cocoon, when finished, is oval, with the longer axis perpendicular, and the larvæ all pupated with the head upwards. They took a week to pupate after the cocoons were completed, and remained in pupa just a fortnight, the female in all cases emerging twenty-four hours sooner than the male. The pupa of *Pl. orichalcea* may be at once distinguished from that of *Pl. gamma*, which is wholly black, by its having the underneath part and the wing-cases of a lovely pale green, a colour which, three or four days before the perfect insects emerge, gradually changes into a dull pink, foreshadowing the colouring of the under side of the abdomen and wings of the imago.

The habit of the larva appears to be to eat the young top leaves of the *Eupatorium*, and work downwards. When not engaged in feeding it rests on the under side of a leaf, grasping the midrib.

One larva, in the course of its last moult, failed to throw off its old skin in its entirety; a narrow band of which remained in an oblique position, embracing the eighth segment, the hinder part of the seventh segment on one side, and the former part of the ninth on the other. This band, as it dried, had tightened, constricting the body till it was only half its normal diameter, and enabling the whole of the internal structure and workings to be plainly seen. As the larva was evidently unable to extricate itself I carefully inserted the eye of a needle beneath the ligature, and, aided by sundry energetic wrigglings of the larva itself, split it asunder. The body soon resumed its usual dimensions, and the larva fed up and turned all right, and the imago emerged apparently perfect; but when I got it on the board I found the left forewing, though not crippled, about one-eighth of an inch shorter than the right.

I think it is quite possible that *Plusia orichalcea* may be more widely spread than is generally supposed. The perfect insect is rarely seen except at night, and the larva would be easily passed over as only *Pl. gamma.* I hope to be able to give a fuller account of its earlier stages another year. (W. Warren, September 17th, 1883; E.M.M., October, 1883, XX, 116.)

PLUSIA BRACTEA.

Plate CII, fig. 4.

I received thirty-six eggs of *Plusia bractea* from Dr. F. Buchanan White on the 1st of August, 1872.

The egg is hemispherical above, but flattened and slightly depressed beneath; the upper surface is finely ribbed and reticulated; below, it is smooth and glistening. Its colour is a greenish yellowish white.

The eggs the day before hatching (on the 7th) became whity-brown, and a blotch at the top appeared, composed of minute brown specks.

The young larvæ were whitish with whity-brown heads, each segment with a transverse row of blackish dots bearing dusky hairs, but these dots and hairs very faintly visible with a strong lens.

The larvæ fed at first and throve well on groundsel up to the end of August, when they had attained the length of three-eighths of an inch, and then they began to show a dislike to their food and to die off. Some were placed on *Lamium maculatum*, which they partook of and looked better, but towards the end of September, they died off by twos and threes, and the last individual died on the 4th of October. (William Buckler, October, 1872; N.B., I, 137.)

April 24th, 1873.—Some of this brood Mr. George Norman, of Forres, had fed on stinging-nettle with success as long as the food was procurable; afterwards, at my suggestion, on *Lamium purpureum*, on which food he succeeded in bringing five or six through the winter. On the above date I received from him the loan of a fine healthy example fed up to the verge of its final moult. At this time it was seven-eighths of an inch long, and stout in proportion, tapering rapidly and considerably from the fourth segment to the head, which is much the smallest segment. Its colour was a pale tint of yellowish-green, becoming rather whitish-green on the hinder dorsal surface; its lines whitish but showing very faintly, two down the middle of the back representing the dorsal line ; these diverge about the middle of each segment, and contract again to a parallel course at the end of it ; on each side of these run two more, the outer one the most distinct, and on these are the faint whitish tubercular dots, each furnished with a fine black hair ; next come a few whitish minute dots along each segment, the ground-colour here growing a little deeper in tint; then comes the spiracular yellowish-whitish line ; on this the flesh-coloured spiracles are situated, and most delicately ovalled with black ; beneath a few faint whitish dots are sprinkled ; the head is of a watery whitish-green, having a black streak down the side of each lobe. The few hairs scattered along the sides are whitish.

At this time it did not eat, and was evidently about to moult, but this operation did not occur until the 29th, when it assumed its new dress of bright green, and the two pale dorsal lines filled up with green rather darker than the ground ; the other pale lines, tubercular warts, and small dots just the same as before, only showing now more distinctly. The head green, with a conspicuous broadish black streak on each side.

On the afternoon of the 30th it began to make its first meal, after fasting at least eight days.

On the 6th of May it was now an inch and a quarter in length, with the spiracles and mouth deep flesh-colour, the former in delicate black oval rings. Hairs blackish towards and on the head, the rest pale flesh-colour ; the tubercles sprinkled on the sides and under the belly ; markings as before ; the inner sides of the anterior legs twice barred with black at the base ; the belly the deepest green. It has preferred *Lamium album*.

This larva continued to feed on *Lamium album* up to the 15th of May, but as from thence to the 17th it had ceased to feed I put it in a jam-pot with food at the bottom, and tied over the top with leno. On the 18th it spun a purse-like web partly on the side of the pot, but chiefly on the leno, and supported below by two leaves of *L. album*, to which it was partly attached; in shape it was horizontally oval, but flattened at the top by the leno, and it measured an inch and three-eighths long by three-quarters of an inch broad, and the same in depth; it was of whitish silk and semi-transparent, so that the larva could be faintly seen within it.

On the 24th I examined the pupa, which was a little over three-quarters of an inch in length, very stout, with the tips of the wing-covers rounded and projecting from the abdomen. Its colour on the head, thorax, and abdomen blackish-brown, the segmental divisions pale pea-green, the wing-cases a bright full green at their edges and extremities, changing gradually from thence towards their base into blackish-brown, and with but little polish.

The moth appeared on the 20th of June, 1873. (William Buckler, June, 1873; N.B., I, 137 and 165.)

On the 3rd of August, 1882, I received from Mrs. Battersby, of Cromlyn, Rathowen, co. Westmeath, a batch of eggs laid by a captured female in a chip box, scattered over the surface singly, and side by side in little groups.

The egg is round or globular, though a little flattened beneath where it adheres to the chip; it is numerously ribbed and reticulated, very slightly glistening, and is of so pale a tint as not to be readily seen on the chip, though on close scrutiny and comparison together the delicate tint of the egg inclines to a greenish straw-colour. On the 4th many showed three brownish but extremely faint small dots, only visible with a strong lens. These dots represented the ocelli and mouth of the head of the embryo, and in the morning of the 5th they began to hatch, and by the evening altogether fourteen or fifteen were disclosed, some much whiter than others, just as about ten or eleven eggs had brown centres, and they will, I suspect, prove to be

not *Plusia bractea*, but another species. They proved
without vitality. Those I took to be *Pl. bractea* were
the most numerous, and their larvæ, when hatched,
were quite white with faintly darker hairs. The
others are rather less white, and have minute black
dots and hairs.

By the 9th of August they all had a watery greenish
tinge, with the internal vessel bright yellowish-green,
showing plainly through the transparent skin.

I fed them at first with groundsel, which relaxed
them so much as to cause a great mortality at the
first moult on the 13th and 14th, when they were from
five to five and a half millimetres long, and the
tubercular black dots on the fifth, sixth, and seventh
segments were larger than those on the other segments. The larvæ were thick behind and tapering
anteriorly.

Eleven I put when hatched on *Stachys sylvatica*
are all doing well at this date. The larvæ are greener
than before, some of a pale yellow-green, others much
darker, showing the dark green internal vessel broadly
through the skin, over which the white lines can be
distinctly traced, and by the 16th of August they
measured seven millimetres in length. On the 23rd
they moulted the second time, and next day measured
from nine to ten millimetres long, when their white
lines were very much more distinct, as well as the
white linear edging to the dark dorsal line. On the
30th some were lying up, and on the 3rd of September
had moulted the third time ; they continued to feed
on *Stachys sylvatica* up to the 11th, and from that
date to the 17th fed but little, and seemed to be getting
torpid ; at this date only one survived of all those
of the first lot that had been so purged by eating
groundsel. By the end of October all had died except
four, yet holding to *Galeobdolon luteum* ; on the 11th
of November only three were alive ; these on the 20th
of that month I placed on potted plants of *Galeobdolon*

luteum; they were about thirteen millimetres long, and much as before.

In 1883 I had more eggs of *Plusia bractea* from Mrs. Battersby, and they hatched early in August, and the larvæ were placed on a potted plant of *Lamium album*, where they kept on the under-side of the leaves, puncturing them with small holes, and eventually a leaf or two had a dissected appearance. Thus they kept out of sight, though their presence was evidenced by larger and larger holes eaten through the leaves. They began to moult the third time on the 9th and 10th, the latest on the 14th, and were then transferred to a fresh-potted plant of *Lamium album*. (William Buckler, August, 1883; Note Book, IV, 151.)

PLUSIA IOTA.

Plate CII, fig. 6.

On the 6th of July, 1874, a few eggs (sixteen) of *Plusia iota* were kindly sent me by Mrs. Hutchinson, of Leominster. They were laid by a large worn captured female, and two or three had already hatched. The eggs were laid singly, adhering to the side of a chip box.

The egg is hemispherical, flattened a little on the centre of its base, numerously ribbed, the ribs radiating from its upper centre, or rather conical apex. At this time it is in colour very pale whity-brown, scarcely to be distinguished from the chip on which they were laid. Around the top of each egg were three brown specks arranged in a triangle. After the extrusion of the larva the shell is left a clear glistening white.

Escaping from the egg by a hole through the side, the larva is then of a brownish whitish gelatinous appearance, or dirty whitish, with tubercular dots of grey-brown; the head with a dark grey-brown mark

on each side and at the mouth, which probably
accounts for the triangular specks on the egg.

In a few days after feeding on *Lamium purpureum*
they became by degrees of a pale yellowish-green, and
from the 1st of August their food was changed to
Lamium album, and by the 1st of September they
measured three-eighths of an inch in length, and by the
16th half an inch in length, and were already hiber-
nating. They were bright yellowish-green in colour,
with slightly darker green dorsal stripe edged with a
whitish-yellow line, followed closely by another such
line, and the subdorsal line similar; the spiracular line
of the same colour, a trifle thicker and more distinct,
the wart-like tubercles whitish and shining, each with
a dark brown hair.

During the winter they moved but little from the
positions taken up on the roof of their cage, but in
January, 1875, they crawled about, and even ate a
little of *Lamium purpureum*. This they repeated at
intervals during March. In April they began to
moult and feed regularly at short periods, eating cow-
parsley, honeysuckle, and *Lamium album*. They
attained their full growth about the 17th of May,
when they measured from an inch and three-eighths
to an inch and a half in length, tapering gradually
from the eighth segment to the fourth, from thence
rapidly to the head, which is narrower than the second
segment and rather flattened. The two hinder segments are very little tapered. The twelfth slopes
downwards from the middle rather abruptly towards
the rounded anal tip. From the tenth forwards the
segments are plump in the middle, and well-defined.
Its colour is a pale and lively, or brilliant yellow-green,
deepest in tint on the sides and belly, where it appears
velvety. The head is rather the darkest green,
broadly marked with black on the side of each cheek,
from the mouth, which circumscribes the top of each
lobe on the crown with a fine black edging. The
triangular piece of the face is finely edged with black,

as are the parts of the mouth. The anterior legs are
also jet-black; the dorsal line is a little darker, rather
bluish-green, narrowing and then widening about the
middle, where it is crossed by a whitish streak, and
from thence it narrows again to the end of each
segment. This is bordered with greenish-white or
whitish (pure white only when younger to half growth)
followed closely by another such bordering, but this is
interrupted by the anterior tubercle on each side, and
at a short distance is an interrupted or broken finer
subdorsal undulating line, broken by the hinder tubercular wart, followed by a few fine scattered specks;
midway between this and the fine clear yellowish-
white spiracular line the green colour of the ground is
deeper like that of the belly and ventral legs; the
spiracles, just beneath this line, are very small, oval,
flesh-colour, delicately outlined with black; along the
sides and on the ventral legs are scattered a few
whitish specks. The whitish borderings to the vas-
cular dorsal line are very soft in character, and seem
almost to melt one into the other. On the thoracic
segments the green ground-colour is a little deeper
than the rest, and has less of the whitish markings on
it; each of the small wart-like tubercles is furnished
with a fine pale hair; the hairs proceeding from the
head are dusky. The antennal papillæ are translucent
green with a double ring of black round the middle;
the ornamentation on the back is whitish-yellow.

The larva spins a cocoon of an inch in length and
five-eighths or even three-quarters of an inch in width,
supported between the stalk and leaves of the food-
plant. The cocoon is very thin, of a pale greyish
dirty whitish tint and semi-transparent, so that the
pupa can be seen within it.

The pupa itself is three-quarters of an inch in
length and rather stout, and on the thorax squarish in
outline, with the head-piece a conical projection
beyond; the wing-cases at their ends and trunk
together form a blunt convex projection low on the

110 PLUSIA IOTA.

abdomen, and with the rounded end of it free from
the abdomen. At first the pupa is green in colour,
irregularly marked with crimson down the back, but
by degrees this changes to blackish, and then the
whole surface becomes black, but not very shining.

The moths appeared on the 4th and 14th of June,
the larvæ having spun up from the 24th to the 30th
of May. One larva, from Mrs. Hutchinson, which
spun up on the 12th of May, produced the moth on the
4th of June. (William Buckler, June, 1875; N.B.,
II, 85, 89, and 107.)

PLUSIA PULCHRINA.

Plate CII, fig. 7.

I have once or twice found the larva of this species
myself, and in different years have received eggs or
larvæ of it from Messrs. J. Gardner, of Hartlepool,
and C. W. Richardson, of Wakefield; but it was not
until 1878 that I had satisfactorily reared it through.
The eggs are deposited in June or early in July, and
are rather small for the size of the moth, round, but
flattened above; the colour very pale dull yellow,
with a few very minute brown dots. They soon
hatch, and the newly emerged larvæ are greyish-white,
indistinctly spotted with black, and the segmental
divisions smoke-colour. They feed on dead-nettle
(*Lamium*) and other low plants until autumn, when
they commence hibernation, having attained the
length of half to five-eighths of an inch. In spring
they recommence feeding, and by the end of the first
week in May are full-grown, and may be described as
follows :

Length about an inch and a quarter, and stout in
proportion; head glossy, with the lobes rounded, and
narrower than the second segment; body cylindrical, and the segments from fourth to twelfth inclusive of nearly uniform size and width; the thirteenth

segment is small and low, which makes the twelfth
have a raised and swollen appearance; from the
fourth to the head each segment becomes considerably smaller than the one behind it, giving the anterior of the larva a very pointed appearance; skin
rough, and, as well as the head, clothed with a few
scattered but rather stiff hairs; the segmental divisions are well-defined; and, like others in the genus,
there are only six prolegs.

Ground-colour of the body and head bright apple-
green; the mandibles, and a rather broad stripe
extending round each cheek, intensely black; dorsal
line darker green than the ground-colour, and edged
on each side with two irregular interrupted white
lines; these lines become confluent on the posterior
segments, and, with the white encircled tubercles,
give the appearance of a somewhat variegated pattern; there is also a very zigzag white line along the
subdorsal region, and a white even line above the
spiracles; spiracles oblong-oval, placed perpendicu-
larly, cream-colour, encircled with brown; the hairs
have the lower part cream-colour, the tips brown.
Ventral surface, legs, and prolegs uniformly apple-
green.

The larva rests with the anterior segments raised
from the food-plant and the back arched like that
of a Geometer, which gives it a rather grotesque
appearance. Like its relatives it spins a moderately
compact white cocoon amongst its food-plant, through
which the black chrysalis can be readily seen. The
moth emerges in about a month—that is, in the
middle of June. (Geo. T. Porritt, Huddersfield,
February 4th, 1881; Entom., March, 1881, XIV, 66.)

On the 6th of May, 1876, I figured a nearly full-
grown larva sent by Mrs. Hutchinson, which she had
reared from the egg on cow-parsley, honeysuckle, etc.,
and sent me on the 25th of April, with another which
died instead of spinning up. But the one which I
figured on the 6th of May spun itself up within a day

or two after, in a light filmy cocoon of silk of a whitish
colour, as large as a pigeon's egg and of similar shape,
spun to the leno cover and partly to the side of its pot.
This cocoon, being very light, was sufficiently transparent to allow first the larva and then the form of
the pupa to be seen plainly within it, and also the old
larval skin next the tail of the pupa.   On opening the
cocoon, the pupa was found to measure rather more
than three-quarters of an inch in length, by nearly a
quarter of an inch in diameter at the thickest part.   It
tapered a little from the thorax to the head, which was
rather produced ; the abdomen was long in proportion,
of about equal size, tapering on the last three segments,
which, at the end, terminate in a rough knob furnished
with a central short spike having two divergent curled
tips, and surrounded with four or five shorter curled-
tipped bristles ; the antenna-cases and trunk, together
with the tips of the wing-covers, are well developed,
and end in a triangular projection rather overlapping
the abdomen.   When opened on the 6th of October,
the moth standing over, its colour was black and without gloss, dull on the wings, the abdomen very slightly
glistening.   (William Buckler, October, 1876 ; Note
Book, II, 135.)

PLUSIA GAMMA.

Plate CII, fig. 8.

*Notes on a probably hitherto undescribed form of the
larva of Plusia gamma.*—In the middle of July, 1892, Mr.
Charles Whitehead, of Maidstone, sent me, on the suggestion of Mr. Stainton, three larvæ for identification.
The note accompanying the larvæ stated they were
abundant, feeding on clover, nettle, thistle, etc., and
that as *Plusia gamma* had been abundant earlier in the
year, Mr. Whitehead thought they must be a form of
the larva of that insect, although so totally unlike the
ordinary well-known form.   On opening the box I saw

at once that Mr. Whitehead was right in supposing
them to be a *Plusia*, but as I had no recollection of
seeing any larvæ of *Pl. gamma* at all like them in
appearance, I doubted the correctness of assigning
them to that species.

Three days later, on July 18th, I received five more
of the larvæ from Mr. Whitehead, and although none
of them were quite so dark in colour as one or two of
the previous specimens, and one was much greener,
they were evidently the same thing. At this time,
too, they began to spin up, although still quite small,
and as the pupæ seemed little more than a third the
size of that of *Pl. gamma*, I became still more confirmed in my doubt about them, and having failed in
my attempts to find a description of any larva agree-
ing with them, I began to have visions of a new
Plusia!

On the 4th of August Mr. Whitehead sent me one
of two moths he had just bred from some of the larvæ,
and on the 6th he forwarded another; whilst, in the
meantime, I also had bred a good specimen, the only
one which emerged from my larvæ. All the moths were
exceedingly small, less than half the size of a number of
ordinary *Pl. gamma* which I netted for comparison
on the Lancashire coast (where the species was flying
in thousands) a fortnight ago. But, apart from size
and the tone of colour, I could find nothing whatever
to distinguish them from *Plusia gamma*. The colour
was very perceptibly paler and more silvery, without
any of the purple tint which characterised all the
freshly-emerged specimens I caught this year, and also
all the specimens in my cabinet.

All the larvæ when full-grown were very small;
probably half of mine died without spinning at all, and
from those that did spin and change to pupæ, only one
moth managed to emerge. Mr. Whitehead also wrote
of his, that "many cocoons were imperfect and came
to nothing."

The subject is very interesting, and possibly some

of the readers of this note may be able to suggest an explanation. The colouring of the larvæ, I am satisfied, was perfectly natural, and not due in any way to disease or feebleness. The larva I described was darker than the majority, but the others were sufficiently near it to be included in the same type.

Length when full-grown about an inch; ground-colour very dark olive-green, in one specimen nearly black; head and prolegs intensely black and shining; two very fine, interrupted, almost inconspicuous yellow lines extend through the dorsal region, followed outside by a broad, bright yellow, double sub-dorsal line, the outer of the two stripes being narrower than the inner; spiracular stripes also broad and bright yellow; tubercles raised, large and distinct, black, surmounted with a pale greyish-yellow spot, though these paler spots are less conspicuous on the side than on the dorsal tubercles; each spot emits a single short stiff hair of the same grey colour; spiracles greyish-yellow, narrowly edged with brown; ventral surface dark olive-green, the prolegs having on the outsides a large cup-shaped black mark. (Geo. T. Porritt, 7th September, 1892; E.M.M., October, 1892, XXVIII, 255.)

The article by Mr. Porritt, on a probably hitherto undescribed form of the larva of *Plusia gamma* immediately brought to my mind a similar experience during last July. On the 6th, 10th, and 14th of that month my boys discovered several small half-looper larvæ whilst searching amongst *Matricaria* for those of *Cucullia chamomillæ*. I was much puzzled with these larvæ, although I had a suspicion that they would produce *Pl. gamma*, but they were so totally different from the usual form that I thought I might be mistaken. Speaking from memory, they exactly coincided with the description given by Mr. Porritt, and it was the dark olive-green colour of the stripes that so especially attracted my attention and caused my doubts, which were increased by the smallness of the larvæ at the time of the pupal change. Only three individuals assumed the perfect state, the remainder of the larvæ having died without spinning up, although constantly supplied with fresh food. Two specimens emerged on the 16th of August, and one later, but all were very diminutive, and had the same pale silvery appearance that struck Mr. Porritt. Two by accident escaped, but the third specimen is now in the cabinet. I may be mistaken, but I fancied that the usual attenuated character of the front segments of the larva of *Pl. gamma* was absent in my individuals. (J. C. Miller, 1st October, 1892; E.M.M., November, 1892, XXVIII, 287.)

PLUSIA INTERROGATIONIS.

Plate CII, fig. 9.

On the 9th of June, 1869, I had the pleasure of receiving the larva, nearly full-fed, of this pretty species from Dr. F. Buchanan White, who had taken several in Inverness-shire, and who during the previous autumn had swept up a few young examples from heather in Ross-shire, four of which he kindly sent to me in October. These were barely one-third of an inch long, and presented the same pattern as the mature larva, being of a full green colour with the subspiracular stripe of sulphur-yellow very conspicuous. They fed occasionally on heather till the end of November, and rested on the stems in a curved posture. Only one, however, lived on to nearly the end of March, and was then half an inch long, and, no young shoots appearing on the heather, it fed a little on a blade of grass and sallow catkin; but one morning I had the mortification of seeing it hang lifeless from a stem.

The full-grown larva measured nearly one inch and a quarter in length when stretched out, though it generally had the anterior half of its body arched upwards, being thick in proportion to its length, tapering gradually from the sixth segment to the head, which is smallest, the hinder segment tapering but little. Viewed sideways the back of the twelfth segment rises a little to the middle, and slopes rapidly downwards from thence to the anal extremity, the two pairs of ventral prolegs being equally developed.

The ground-colour is a bright and deep full green, but paler on the back, though the dorsal stripe is as dark as the sides, and begins wide, narrows, then swells wider to an angle in the middle, decreases similarly, and widens towards the end, and is finally edged throughout with greenish-white. This is its course through all the segments except the thoracic, where it is more simple and linear.

The subdorsal line is greenish-white, finely edged with darker green, and midway between the dorsal and subdorsal is a tortuous line of greenish-white, on which are the usual tubercular warts of the same colour, each bearing a fine brown hair. The sub-spiracular stripe is sulphur-yellow, and the belly and legs are not quite so green as the space between the subspiracular and subdorsal lines.

The head is green, finely freckled with greenish-white, and having a black streak round the sides to the mouth. Some very small yellow scattered tubercles on the ventral surface.

On the 11th June it began to spin its pale grey silken oval cocoon amongst the stems of heather, and a few days later the pupa became dimly visible through it, lying in the middle in a nearly horizontal position, the head being lowest; its length about half an inch, the wing-cases nearly as long, their tips uppermost, and projecting in a blunt point; from them the abdomen is bent downward at a right angle, having a blunt anal point attached to the shrivelled coat-skin of the larva; its general appearance rounded, obtuse, and thick, of a blackish-brown colour, and with scarcely any polish.

The perfect insect emerged on July 8th, 1869. (W. B., 1869; E.M.M., August, 1869, VI, 65.)

TOXOCAMPA CRACCÆ.

Plate CIV, fig. 2.

Larva (when full-grown) one inch and a quarter to one inch and a half in length. When viewed from above it tapers towards the head, and still more towards the posterior end; but when seen sideways appears almost uniformly long and slender.

Its manner of walking is a partial looping of the first six segments; the first two ventral prolegs are very slightly shorter than the others, but to such an extent as to be scarcely noticeable, and it generally assumes an undulating posture in repose along the stem of its food-plant.

Along the back, commencing on the head, is a dark brown, very finely mottled, broad stripe, widest along the middle segment, and with an additional widening on the eleventh segment; in the centre of this is a thin, rather paler stripe, enclosing the very dark brown dorsal line. The subdorsal stripes are double, brown, with a paler ochreous-brown ground, followed by a pale stripe of ochreous, enclosing a very thin brown line; the lateral lines double, dark brown, extending from the mouth to the anal prolegs; edged above with black at the anterior portion of each segment; the upper one widening below in the middle, along which are some black dots. Belly and legs brown. Within the dark portion of the back, on each segment, are placed four black dots in the usual order, and on the eleventh segment there is an additional black dot on each side, outside the dark region. Subdorsal lines also containing two black dots and a minute ring.

Went to earth on the 24th of June. (W. B., July, 1865; E.M.M., August, 1865, II, 67.)

On the 4th of May I received, through the kindness of the Rev. Mr. Horton, three little larvæ of this species. They were from a quarter to half an inch in length, and of a slender figure, using only two pairs of their ventral feet.

One of these larvæ soon perished from a bite he had received during his journey from one of his companions; and another died not long afterwards, as I believe, from my neglecting to supply it with *young tender* shoots of vetch; whilst the third, after dwindling for a while, soon recovered its health when furnished with food tender enough for its taste.

As I could not procure *Vicia sylvatica*, Dr. Knaggs told me that *Orobus tuberosus* and *Vicia sepium* would replace it, but I found that the young shoots of the last-named plant were most approved of. (John Hellins, July, 1865; E.M.M., August, 1865, II, 68.)

STILBIA ANOMALA.

Plate CIV, fig. 3.

On the 13th March, 1879, I received two very distinct forms of the larva of this insect from Mr. G. C. Bignell, of Stonehouse, Plymouth, to whom they had been sent from Torquay.

Length about an inch, and of proportionate bulk; nearly uniformly cylindrical; head rounded and polished, about the same width as the second segment; segmental divisions well defined; skin soft and smooth, but not glossy.

Var. 1 has the ground-colour a warm pale chestnut-brown; head greyish-brown, thickly freckled with dark brown; two purplish-brown lines (black at the divisions of the hinder segments), enclosing a yellow line between them, form the dorsal stripe; subdorsal stripes yellow, very finely edged with a darker shade of brown than the ground-colour; spiracular stripes greyish-white, edged above with smoke-colour; spiracles black, those on the second and twelfth segments very large and distinct. Ventral surface, legs, and prolegs uniformly dingy chestnut-brown.

Var. 2 has the ground-colour bright pea-green, with just a tinge of yellow; head of the same colour, but thickly freckled with brown; two lines of a darker green than the ground, enclosing between them a white line, form the dorsal stripe; subdorsal stripes white, finely edged with a darker green than the ground-colour; spiracular stripes white, edged above with smoke-colour; spiracles black as in var. 1. Ventral surface, legs, and prolegs uniformly of the same colour as the ground of the dorsal area.

Feeds on grass. (Geo. T. Porritt, December 2nd, 1879; E.M.M., February, 1880, XVI, 210.)

CATOCALA FRAXINI.

Plate CIV, fig. 4.

On the 2nd of July, 1886, I received two larvæ of *Catocala fraxini* from Mr. R. C. Ivy, of Southport. One of them was nearly adult, the other about half-grown. The former I described as follows :

Length nearly three inches, but slender in proportion. Head broader than the second, but a little narrower than the third, segment, flattened in front and slightly notched on the crown. Body of fairly uniform width, but the seventh, eighth, ninth, and tenth segments the widest; it is round above, but flat ventrally; there is a small hump at the back of the ninth segment, and a smaller ridge at the back of the twelfth segment; segmental divisions clearly defined; skin smooth and without hairs dorsally, but there is a row of tolerably dense short hairs pointing downwards below the spiracles, dividing the dorsal from the ventral area.

The ground of the dorsal area is putty-colour, with strong greenish tinge, and freckled, particularly at the hinder part of each segment, with minute brown dots; head pale pink, surrounded at the back with a conspicuous band of dark damson-plum colour; this band narrows off to a point on each side the face; the upper part of the face is also reticulated with this colour, but towards the mandibles are several dark brown streaks, whilst on each mandible and also on each side is a dark brown spot; dorsal stripe very narrow, green; spiracles oblong-oval, black, encircled with greyish-white; the hump on the ninth segment is darker than the ground-colour, the dark colour extending backwards, and forming a somewhat horseshoe-shaped mark; the back of the ridge on the twelfth segment is also of this dark colour; segmental divisions of the same pink colour as the head.

Ventral surface very pale greenish-white, with a large and conspicuous, nearly triangular, almost black mark on the seventh, eighth, ninth, tenth, and eleventh segments, and there are paler, more rust-coloured marks on the centre of the third and fourth segments; legs and prolegs of the same colour as the ground of the ventral area, the anterior ones being tinged on the outside with pink; hairs greyish-white.

In the half-grown larva the head is considerably wider than any of the following segments, and the colours generally are of a darker shade all through. The head is of a darker pink, but this colour is nearly lost in the broad, dull, black band at the back, and the greater amount of equally dark reticulation on the face; the ground on the dorsal area is much browner, and the narrow dorsal line is almost black; the tubercles, which are not noticeable in the adult larva, are distinct, ochreous-yellow; the spiracles are not so dark, and consequently much less conspicuous; whilst the ventral surface is pinkish-grey, and the outside of both anterior legs and prolegs, together with the hairs, pink.

The species feeds on ash and poplar, and both young and adult larvæ rest at full length on the twigs or small branches ; the six anterior legs, and the pro-legs on the ninth, tenth, and ventral segments, which are larger than the others, are spread out from the body, and give the larva a very sprawling appearance.

Both larvæ spun loose cocoons among the dead leaves, etc., at the bottom of their cage, and changed to ordinary-shaped pupæ of a purple-plum colour, powdered with greyish as a ripe plum also is.

The moths, two fine specimens, emerged on August 25th and 26th, 1886, respectively. (Geo. T. Porritt, January 8th, 1890 ; E.M.M., May, 1890, XXVI, 125.)

CATOCALA NUPTA.

Plate CIV, fig. 5.

I received eggs during the winter (1870–71) from Mrs. Hutchinson. The egg is circular, rounded and convex above, rather flattened beneath, and ribbed. It is of a brownish-grey colour, with two zones of blackish encircling them, separated only by a narrow ring rather paler than the ground-colour ; the ribs themselves of the grey ground like the central patch at the apex. (William Buckler, 1871 ; Note Book I, 67.)

CATOCALA PROMISSA.

Plate CV, fig. 1.

On August 26th, 1875, Mr. J. Ross, of Bathampton, most kindly sent me thirty-nine eggs of this species, being the whole produce from six imprisoned female moths captured by him in the New Forest on August 2nd, and with them the permission to select some for myself.

The eggs had been laid from the 9th to the 16th of the month, some on oak bark, the others extruded through the interstices of the leno covering of their

cage, to which they adhered; they were of two different colours, and I contented myself with choosing
three of each, and returning the remainder to Mr.
Ross, from whom I afterwards heard they all proved
sterile.

The egg of *Catocala promissa* is of a good size, of a
rather flattened spherical figure, a little depressed in
the upper centre and much more beneath, the shell
covered with coarse, projecting, sinuated ribs, varying
from fourteen to eighteen in number, so close together
as almost to hide the surface between them, the depressed spot in the centre of the top coarsely reticulated; when fertile it is of a dull drab colour, and so
continues through the winter, but, as I found, when
sterile it is dark brown, and eventually shrivels up.

About the middle of April, 1876, while looking at
the three drab-coloured eggs, I fancied two of them
seemed rather more plump than before, and a close
examination proved this to be the case, as a little of
the smooth shell had become visible between the rough
ribs, and the upper central hollow nearly filled up;
this last on the 18th was completely rounded over,
and the ribs were turning paler; on the 20th they had
become whitish, and the interstices greenish-drab
colour; and on the morning of the 21st I found one
larva was hatched; the dull, whitish, empty shell
showed a large hole in the side, through which the
larva had escaped; the next morning I saw a second
had hatched. At this time none of the oak buds had
burst, nor were many much swollen, but I picked open
two or three at a time of the best to be found for the
young larvæ to feed on, Mr. Ross also, at this
juncture, kindly supplying me with a few tender oak
leaves which he had contrived to force out; but in
placing this food in the cage I noticed one of the
larvæ, when put on the leaves, swing away from them
by a thread, and though I replaced it before shutting
the cage, yet it must have again swung out, for at that
moment I unconsciously lost it; however, next morn-

ing (the 25th) I was somewhat consoled at seeing the
third was hatched, and so I again had two young
larvæ to watch. Curiously enough, neither of them
seemed to care then for the leaves, but chose the buds
and those containing blossoms in preference, feeding
only after dark, and resting all day stretched out at
full length, motionless, belly upwards on the muslin
cover of the cage, a habit continued through all stages
of growth, the moulting included, a process which invariably occurred at night, in that position, as proved
by the cast skin next morning adhering to the muslin
with all the legs spread out to their full extent.

No doubt, in a state of nature, the larva passes the
daylight in this quiescent position, probably on the
under surface of horizontal or sloping twigs or
branches of the oak, where it would be in shadow, and
would assimilate wonderfully well to the more or less
lichen-covered surface on which it would be closely
pressed, and would be in a great measure safe from
the prying eyes of birds, and, I may add, of entomologists, for I remember no recorded instance of its
having been found at large by any who have collected
in that favourite hunting-ground, the New Forest.

The newly-hatched larva was three-sixteenths of an
inch long, with largish head and slender body, stoutest
at the ninth and tenth segments, the first two pairs of
ventral legs quite rudimentary, the third and fourth
pairs conspicuously developed, and also the anal pair ;
its mode of progression was precisely similar to that of
a geometer ; the colour of the head black, of the body
a light drab, broadly banded with dark brownish-grey
across most of the segments, with fine pale double longitudinal lines along the sides, and with two pairs of
black dots and bristles on the back of each segment ;
after the first moult the dark bands disappeared, and
the colouring was light greenish-grey, the dorsal line
showed as darker, and then a lighter spear-shaped mark
on each segment ; the pale twin-like subdorsal lines
still remained, and below them a blackish blotch on the

side of each segment ; after the second moult, at
the end of a fortnight, the larva was five-eighths of
an inch in length, and of stouter character, having an
elevated ridge on the back of the ninth and twelfth segments, the anterior pairs of ventral legs now first in
use for walking over the food by night ; the colouring very lichenous in appearance, no lines on the
sides, but large and conspicuous whitish blotches on
the fifth, eighth, and ninth segments, the elevated
ridge darker grey than the rest ; in another week,
when the length of seven-eighths of an inch was
attained, a whitish narrow streak appeared over the
crown of the head, and the ridge on the ninth segment
became black, the rest of the body light greenish-
grey with paler blotches as before; on the 16th of
May one of the two larvæ fixed itself for a moult, but
died on the 19th, unable to complete the operation.
Meanwhile the remaining larva throve well, and by the
21st had become one inch and three-eighths in length,
the growth being rapid now, the colouring much as
before, very lichenous in appearance; the last moult
occurred during the night of the 23rd, and the next
morning I found it measure one inch and three-
quarters in length, the general colouring a rather
greener grey than at any previous stage ; even the
whitish blotches were now faintly tinged with greenish-
ochreous ; on the 26th it had reached its full growth,
when I took its third portrait, and a full description
which follows presently ; on the 28th it was shorten-
ing evidently, although continuing to feed at night till
the 30th, when it had decreased considerably, and was
irritable at the least disturbance, and on the 31st it retired amidst some sprays of oak, and entered a little
way into some light soil beneath, where it formed a
cocoon composed chiefly of small particles of dry stalks
and roots with peat earth, and lined, as I afterwards
found, with coarse whitish silk, disposed in very large
meshes, yet smooth enough ; the upper surface being
just level with the surrounding soil, and partly attached

to a stone I had placed there. The moth, a female, appeared on the 24th of July.

The full-grown larva is two inches and one-eighth in length, the body thickest at the ninth and tenth segments, tapering from thence a little gradually to the head, and a little more to the anal extremity ; the head rises a little on the crown, where the lobes are slightly defined, and is flattish in front ; there is a prominent ridge having a triangular hump on the back of the ninth segment, and a slight elevation occurs near the end of the twelfth, bearing the hinder pair of tubercles more sharply prominent than the rest ; the back is rounded, the belly flattened ; at the junction of the two surfaces just above the legs is a fringe of fleshy filaments, more or less branched, though a few simple ones occur amongst them ; the anterior pairs of tubercular warts on the back are small and unobtrusive, while the hinder pairs, and the single row along each side, are rather large and bluntly pyramidal, every one having a fine bristle ; the anterior and ventral legs extend laterally at right angles to the body, the anal pair also at an obtuse angle backwards, the third ventral pair long, and the fourth pair longest. The ground-colour is a light greenish-grey, with a distinct, large, pale patch of faint ochreous-greenish on the sides and back of the fifth, another on the ninth, and on the tenth less and less pale, strongly contrasted towards the division by a sooty transverse irregular band extending down either side from the blackish hump on the ninth to the back of the leg, from whence it spreads behind, at first broadly, then slants off to a point on the lower side of the tenth ; the end of the twelfth segment is a little darkened ; the head is light greenish-grey, reticulated with darker grey ; a transverse streak of black reticulation over the crown extends to the mouth, defining the boundary of the face ; behind this a shorter black streak marks the back of the cheeks ; the face itself is whitish, with a dark greyish streak on either side downwards to the mouth. The thoracic segments are very much covered with freckles of lightish grey, dark grey, and black, some of them so disposed as to faintly indicate dorsal and subdorsal double lines ; on the fifth the back, though pale in front, is clouded behind, while on the sixth, seventh, eighth, and all beyond the ninth it is rather uniformly covered with fine greenish-grey freckles, forming on each somewhat of a truncated diamond shape, each successively growing paler ; from the sixth to the end of the eighth these diamonds are relieved by the hinder pairs of whitish prominent warts, more or less ringed at their base with dark grey or black ; from these proceed backward to the segmental division short, dark greyish, double lines rather convergent, most strongly defined on the fifth, eleventh, twelfth, and thirteenth ; along the sides, from the end of each segment, is a broad-based, somewhat wedge shape of the paler ground, flanked below by the lateral whitish wart, from whence a pale sinuous streak ascends a little obliquely forwards, finely and sharply edged below with black like the wart itself ; the dull red oval spiracle, outlined with black, comes close beneath in front of the wart ; the rest of the side is freckled to about the same appearance as the back ; the fleshy filaments are pearly white ; the anterior legs pale, and ringed with dark greenish-grey ; the two first ventral pairs are whitish-grey, the third and fourth pairs greyish in front, darker greenish-grey behind, bearing a few black freckles, the anal pair similar ; the belly is whitish, with a conspicuous blackish mark on the middle of each segment, viz. a transverse bar between each pair of the anterior legs, a largish round spot on the fifth and sixth segments, a very much larger spot on the seventh, eighth, ninth, and tenth ; on these two last they are elongated transversely to a diamond shape ; the spot is round on the eleventh, twelfth, and thirteenth, each smaller in the order mentioned ; the skin of the head, back, and sides a little rough, the belly smoother, the filaments smooth.

The pupa is nearly an inch in length by five-sixteenths in diameter across the thorax, which is rounded and sloping smoothly to the head in a convex curve; behind, on the back of the abdomen, is a slight depression, the wing-covers smooth, and from them the abdomen is full, but soon tapers rather sharply to the tip, which is rough and furnished with several converging, curled-topped spines; the colour of the skin is purplish-brown, the abdominal divisions dingy red, though this local colouring can only be seen on parts that happen to be rubbed, as the surface generally is covered with a fine, opaque, powdery, bluish bloom; a few short, fine, light brown, bristly hairs, pointing behind, are sparingly distributed over the abdomen. (W. B., December 2nd, 1876; E.M.M., XIII, 233, March, 1877.)

CATOCALA SPONSA.

Plate CV, fig. 2.

In August, 1865, I captured at sugar a moth of this species, which proved to be a female, and she obligingly laid a few eggs on oak twigs and the sides and leno cover of her cage, after being fed for a fortnight with moistened sugar.

The eggs were circular and rather depressed, smooth and shining, olive-brown, some of them semi-transparent and mottled with darker brown, showing a whitish ring near the margin and a narrow blackish ring within it; these last, as the sequel proved, were fertile, and the others barren.

In April, 1866, the young larvæ hatched just as the oak buds and blossoms began to appear, and on which they fed, preferring the blossoms, though, after their second moult, they readily partook of the leaves.

When first hatched they were blackish-brown, with a few paler blotches, long in proportion, looping with much activity in their progression, often standing erect on their anal legs with a tremulous motion of the body, and, if touched, falling and wriggling in an excited manner.

After the second moult they were of a very pale brown mottled with olive-greenish and brown, exhibiting decidedly the peculiarities of contour pertaining to larvæ of the genus *Catocala*.

In their early stages they were very restless for some time after being disturbed by changing their oak twigs, walking about their glass prison as if bent on escaping, but would at length settle down to their food; in repose they were generally stretched out close to the surface of the twigs, and assimilated well with them; as their size increased, so, in proportion, they became quieter, and at length even lethargic in their demeanour, each individual having a separate residence. On arriving at maturity they spun a loose kind of hammock amongst the oak leaves, and therein changed to pupæ of a purplish-red colour, covered with a delicate violet bloom. The moths appeared towards the end of July.

The full-grown larva, when stretched out, measures two inches or two inches and a half in length; its walk is a half-looping motion, sometimes retaining that posture in feeding, though generally it closely embraces the twig, its body being extended and its head erected to the edge of a leaf.

In form it is rounded above and flattened beneath, and tapering towards each extremity. The head is broad, rounded, slightly elevated, and indented on the crown, and is a trifle larger than the second segment. There is a transverse dorsal hump on the ninth segment, and the twelfth also appears slightly humped, but the thirteenth is much depressed. The thoracic segments are deeply wrinkled, the others plump and deeply indented at the divisions.

Tubercles conical, and terminating in a very short spiky bristle, six on each segment, viz. two lateral and four dorsal, the hinder dorsal pairs being much the largest.

In colour the head is of a deep dull red, brighter on the face; the edge of the crown bordered with black, and edged beneath in the centre with pale ochreous, and on each side, just below this, a black spot. The thoracic segments much suffused with greyish-brown; tubercles and markings rather indistinct.

The ground-colour of the body is pale ochreous, pale brownish, or greyish-ochreous; a large bright pale ochreous patch on the fifth segment, occupying its anterior dorsal surface, and extending a little down each side of it. There is a purplish-brown or grey blotch transversely suffused on the hinder part of the ninth and beginning of the tenth segments, and following on the latter a paler patch. A similar dark blotch on the chief portion anteriorly of the twelfth segment. The dorsal and subdorsal stripes purplish-brown or brownish-grey, with a narrow line of pale ground-colour between them; the subdorsal stripes have the tubercles placed thereon, and the stripes widen round the bases of the tubercles. There are two similar lateral stripes, the lowest not very distinct, owing to the aggregation of dark atoms along the sides, all the stripes being composed of minute spots.

Sometimes faint indications occur of greyish transverse bands on the sixth, seventh, and eighth segments. Tubercles deep glossy red posteriorly and black anteriorly, but sometimes all are black except those on the twelfth segment. Spiracles dirty whitish or brownish, margined with dark brown. Filaments pinkish-grey. The ridge of the hump on the ninth segment has generally a very pale blotch of the ground-colour, divided by a narrow transverse black mark between the tubercles. Belly pale greyish with dark red spots. (W. B., 1867; E.M.M., May, 1867, III, 276.)

EUCLIDIA MI.

Plate CV, fig. 3.

A moth of this species, taken on the 5th of June,
1886, at the Green Farm Wood, Doncaster, deposited
eggs which were globular in shape, the colour a dull
pale green. They hatched about the 28th of the same
month, and the young larvæ were dingy green with
large yellowish-brown heads; when walking they looped
the back in the same way as does a geometer, and
when disturbed at once rolled themselves up and
feigned death. They fed well on grass and common
white clover, and by the 22nd of July were slender
creatures of about five-eighths of an inch long, with
only six ventral legs, and consequently were veritable "loopers," arching the back as much as any
geometer. On the 7th of August, when they were
almost an inch long, I described them as follows :

Very slender ; head wider and deeper than the second
segment, the lobes evenly rounded ; body of nearly
uniform width throughout, rounded above, slightly
flattened ventrally ; skin smooth, the segmental divisions clearly defined, but not deeply cut ; there are
only three pairs of ventral legs, on the ninth, tenth,
and thirteenth segments respectively ; the last pair,
when at rest, being stretched backwards and outwards,
give the appearance of a notched anal prominence.
Ground-colour generally dingy pale olive-green, in
some specimens, however, bright greenish-yellow ; on
it is a pretty ornamentation of chocolate-brown stripes
as follows :—First, a narrow and interrupted medio-
dorsal, then a double and more clearly defined one,
followed below at about the same distance by another
double stripe ; then follows a broader one, and immediately adjoining it is the broad and conspicuous
lemon-coloured spiracular stripe ; all these stripes
extend in strong relief through the head. The ventral
surface has a somewhat similar but not so clearly

defined ornamentation; the ground-colour being as
on the dorsal area, and having a central, then a
double, followed by another double line, all of chocolate-brown.

The habits, perfect manner of looping the back
when walking, feeding, and everything else are exactly as in a geometer; and were it not for the
additional pair of ventral legs it would be impossible
to separate it from a geometer.

The adult larva I described on the 6th of September
as follows:

Length about one and a half inches, and, although
still rather slender, is considerably stouter in proportion than when last described; head larger and slightly
wider than the second segment; body cylindrical above,
a little flatter ventrally; it is of nearly uniform width
to the ninth segment, but this and the tenth are a
little wider; from the tenth it tapers more strongly
to the anal extremity; skin smooth, the segmental
divisions clearly defined but not deeply cut, and there
is a very slight rather puckered ridge along the
spiracular region; there are only three pairs of ventral
legs, on the ninth, tenth, and thirteenth segments
respectively; the last pair, when at rest, are stretched
backwards and outwards, and give the appearance
of being a notched anal prominence. Ground-colour
bright lemon-yellow, some specimens having an ochreous tinge; the same pretty ornamentation of stripes
still remains, as follows:—First, a fine double and
rather irregular pale chocolate-coloured medio-dorsal,
followed by two broader and darker chocolate, then
two more equally broad ones of the paler chocolate,
followed by a narrow one of a darker shade of the
same colour, and closely followed by a still darker one
immediately edging the pale, bright, lemon-yellow,
broad, spiracular stripe; these stripes extend in strong
relief through the head down to the mandibles, and
the whole ornamentation, taken with the ground-
colour, forms a series of alternately chocolate-brown

and lemon-yellow stripes. The ventral area is less distinctly marked than when last described; the ground is greyish-yellow in the centre, rust-colour at the sides, with double interrupted chocolate central stripe; at the sides are two other similarly coloured stripes, the outer edge of the last being close to the broad spiracular stripe; legs and prolegs greyish-yellow, the latter marked on the outside with rust-colour.

Manner of feeding, walking, etc., just as when last described. The last two larvæ went down on the 21st of September, but no imagines afterwards emerged from any of them. (Geo. T. Porritt, May 9th, 1888; E.M.M., June, 1888, XXV, 13.)

EUCLIDIA GLYPHICA.

Plate CV, fig. 4.

On July 2nd, 1878, I received a good supply of eggs, together with the parent moth, of this species from Mr. Blackall of Folkestone.

The eggs were globular, and distinctly ribbed from the summit to the base; when first deposited they were bright pea-green, but soon changed to dull green, with, on the crown, a large brown blotch, and below this blotch a ring of the same colour. They began to hatch on the 10th of the same month, but the young larvæ were not all out before the 13th.

The newly-emerged caterpillar looks large for the size of the egg, being about three-sixteenths of an inch long, is very lively, and when walking arches its back like that of a geometer. Colour a dingy semi-transparent pale green, barred with dark brown, or nearly black; head pale wainscot-brown and polished; and there are rather long hairs scattered over the body.

They fed up well and rapidly on both the white and red species of clover, and when from an inch to an inch and a quarter in length I described them as follows :

Long and slender for the size of the moth ; body evenly rounded above, flattened below, tapering a little at the extremities ; the head has the lobes rounded, and is a little wider than the second segment; skin smooth but not polished ; segmental divisions well defined ; the anal prolegs extend beyond the fold, and form a distinct angle. By this time they have lost the true looper style of walking, but are still half-loopers, having no prolegs on the seventh, eighth, eleventh, and twelfth segments. The ground-colour varies from pale salmon to dull pink, some specimens having a strong yellowish tinge ; a distinct double yellow line, enclosing another very fine still paler line, forms the dorsal stripe ; the anterior point of the pale line on the crown of the head forms the apex of a triangular mark, the base of which is over the mandibles ; the rest of the head is very dark brown ; the subdorsal lines are dull bluish, bordered with smoke-colour, and enclose fine pale greyish lines ; below the spiracular stripe is another irregular greyish line, and below this, but above the spiracles, is another line of pale bluish, edged with smoke-colour ; the spiracular stripes are yellow, rust-colour, or pink, in different specimens. The colours, indeed, vary considerably in different examples, in some the blue side-stripes being scarcely discernible ; spiracles black, as are also the tubercular dots, which, though small, can be distinctly seen with a lens. Ventral surface dull dark smoky-purple, with two yellow central lines.

Most of the larvæ were full-grown by August 7th. Length an inch and three quarters, and the salmon and pink colours of the younger specimens altogether lost. The ground is now of various shades of ochreous-yellow, the darker specimens having a strong rust tinge along the sides ; head of various shades of brown, in some being of a dark sienna colour ; in all there is the pale yellow front triangular mark so noticeable in the earlier stage, and there is also another distinct streak of yellow on the side of each lobe; a brown stripe enclosing a very fine yellow line, and broadly edged outwardly with yellow, forms the dorsal stripe; a double smoke-coloured line composes the subdorsal stripe, and between it and the dorsal stripe are two other irregular yellow lines; above the spiracles is a yellow line edged on each side with smoke-colour, and between it and the subdorsal stripe another irregular yellow line; spiracles and tubercular dots black.

Ventral surface of various shades of dull ochreous, with two greyish central lines; a black mark on the seventh and eighth segments, and a smoke-coloured stripe below the spiracles.

Feeds during the night; in the daytime remains extended at full length, flat along the stalks of the food-plant.

The cocoon is composed of bits of the food-plant firmly knitted together with very closely woven silk; in a state of nature, however, it would probably be on the ground.

The pupa is about five-eighths of an inch long, and of the ordinary shape, though rather blunt and dumpy; colour deep purplish-brown, with the abdominal divisions and spiracles still darker; it is powdered over with a very pretty violet bloom, though more so on the head, thorax, and wing-cases than elsewhere.

From these larvæ I reared a long and beautiful series of imagos the following June. (Geo. T. Porritt, January 8th, 1881; E.M.M., February, 1881, XVII, 210.)

PHYTOMETRA ÆNEA.

Plate CV, fig. 5.

On July 30th, 1865, I received a batch of eggs from Dr. Knaggs. These were of the ordinary *Noctua* form,

rounded, ribbed, and with flat under-side ; in colour
they were by that time a dull purplish-brown. On
August 5th the larvæ began to emerge, little translucent almost colourless loopers to look at, but luckily
they did not all come out at once, otherwise I should
have had little to say about them ; I tried them with
every plant I could think of, but at first with no
success, and by far the greater part of them had died
of starvation, when luckily it came into my mind that
the very last specimen of the moth which I had captured myself was flying over or near some plants of
milkwort (*Polygala vulgaris*) ; a little bit of this plant,
therefore, was put in amongst the other twigs and
leaves, and in a short time, to my great delight, the
five surviving larvæ had all found it out, and were
eating it very freely. They soon began to show an
increase in size, and turned pale green in colour, and
although looping very much, it was easy to see that
they had two pairs of ventral legs. I noticed that
whilst they were small their tint depended on the part
of the plant they ate, the *blue* flowers (I could find no
pink ones) causing them to appear of a dark bluish-
green.

In the first week of September they attained their
full growth, and were then an inch long, slender if
compared with other *Noctua* larvæ, but moderately
stout for loopers ; uniform in width when viewed from
above, but when seen sideways cylindrical in the
middle segments, and flatter towards the head and
tail ; the skin smooth ; the head round ; legs twelve,
the ventral pairs being on segments nine and ten, and
rudiments of another pair, too small for use, on the
eighth.

The colour is a velvety full green, scarcely paler on
the belly ; the head mottled with faint brown ; a hasty
inspection would scarcely detect any lines, but on
looking closely the dorsal vessel appears as a darker
green thread, bordered with paler lines, between which
and the spiracles come three pale subdorsal lines ; the

spiracles yellowish, below them a broader pale line,
which on segments ten to thirteen becomes whitish.
The segmental folds yellow, the usual dots very small,
black, surrounded with light rings, and emitting small
bristles.

When full-grown their walk is semi-looping, and
they rest extended straight and flat on the stems of
their food ; if disturbed they drop off, and fling themselves about angrily.   About September 10th they
began to contract in length and to grow pale, and in
a day or two spun themselves up in very tight-fitting
little cocoons of close-woven grey silk, wrapped about
with some of the leaves and stems of their food.

I have no doubt that in this case it is no *substitute*
food, but the natural one which I have been lucky
enough to discover.   (John Hellins, October 5th, 1865,
E.M.M., December, 1865, II, 163.)

At p. 163 of the second volume of the ' Ent. Mo.
Mag. ' I gave a description of the larva of this species,
with some notes on the egg and cocoon, which I wish
now to amend and enlarge.

On June 19th, 1873, I obtained three or four eggs
from a moth which I had shut up in a glass cylinder
with sprigs of milkwort ; however, she chose to deposit only on the leno covering.   On examining these
eggs with an inch object-glass, I found that they did
not correspond with the short description I had given
in 1865, and fancied I had somehow got hold of
another species, but in due time the larva appeared,
and looked and behaved so exactly like the former
brood, that I became quite satisfied that my puzzle
arose from my not having examined the eggs formerly
so minutely as I had now done.   They had then come
to me not long before the hatching of the larvæ ; I
must have looked at them with a lens of low power,
and so missed their true structure.

I find the egg, then, is of the usual *Noctua* shape,
somewhat flattened, the apex occupied by a small round
patch of tiny irregular network ; all the rest of the

shell, down to the flat under surface, covered with a most beautifully regular three-cornered reticulation, so exactly designed that, wherever the eye rests, it involuntarily forms hexagons out of half-dozens of the triangles. Each of the knots at the angles of the network is furnished with a comparatively longish curved spine; the colour of the shell is whitish, mottled with long blotches of pale pink, which are disposed horizontally round the egg; the lines of triangular network are pink, the spines pink with brown tips. A short time before the larva is hatched the egg becomes purplish all over.

To the present date this is the most remarkable egg I have seen, and whilst contemplating its spiny ornamentation one cannot help being reminded of old Gilbert White's remarks on the parturition of hedgehogs! (Letter xxxi, to Thos. Pennant, Esq.).

Of the larva I have nothing fresh now to say.

The cocoon, made of a tough texture of greyish-white silk, is not quite half an inch long, and about three-sixteenths of an inch wide, with a few outside threads to draw round the surrounding leaves, etc. The pupa is nearly three-eighths of an inch long, cylindrical, slender, and remarkably even in bulk throughout (reminding one in this respect of the pupa of an *Hepialus*), blunt at the head, the abdominal rings deeply cut, the last segment alone tapering, and ending in a blunt tip with two extremely short blunt spikes; the colour on the head and wing-cases a rich olive-tinted brown, on the rest of the body a bright reddish-brown; the skin rather glossy. (John Hellins, September 20th, 1873; E.M.M., November, 1873, X, 139.)

The following list of parasites, bred from the larvæ
or pupæ of the species included in the present volume,
has been kindly prepared by Mr. G. C. Bignell,
F.E.S.—G. T. P.

HOST.	PARASITE.	By whom bred.
Dianthæcia irregularis	*Ichneumon xanthorius* Forster	Mrs. Hutchinson.
	Ophion distans Thoms.	G. C. Bignell.
	Limneria ruficincta Grav. ..	Mrs. Hutchinson.
	Apanteles sericeus Nees	Mrs. Hutchinson.
D. carpophaga	*Ichneumon bisignatus* Grav.	C. S. Gregson.
	Amblyteles castigator Fab. ...	G. Elisha.
D. capsincola	*Eurylabus tristis* Grav.	B. A. Bower.
	Cryptus obscurus Grav.	H. D'Orville.
	Ophion luteum L. {	G. C. Bignell. / H. W. Barker. / T. A. Marshall. / —. Neville.
	Agrypon tenuicorne Grav. ...	G. C. Bignell.
	Paniscus testaceus Grav.	H. W. Barker.
	P. cephalotes Holmg.	H. W. Barker.
	Apanteles sericeus Nees	G. C. Bignell.
	Microplitis spectabilis Halid.	H. D'Orville.
	M. tristis Nees	W. West.
D. cucubali	*Hemiteles furcatus* Tasch. ...	G. C. Bignell.
	Ophion luteum L.	T. A. Marshall.
	Limneria ruficincta Grav. ...	G. C. Bignell.
	Pimpla brevicornis Grav.	G. C. Bignell.
	Apanteles sericeus Nees..	G. C. Bignell.
	Microplitis tristis Nees	G. C. Bignell.
Hecatera serena	*Henicospilus merdarius* Grav.	B. A. Bower.
	Sagaritis zonata Grav.	G. C. Bignell.
	Limneria ruficincta Grav. {	G. C. Bignell. / J. P. Cregoe.
Polia chi	†*Microplitis mediana* Ruthe ...	Miss N. P. Decie.
P. flavocincta	*Microplitis mediana* Ruthe ...	G. C. Bignell.
	Exorista vulgaris Fallén	G. C. Bignell.
	E. parens Reinh.	G. C. Bignell.
Epunda viminalis ...	*Limneria exareolata* Ratz. ...	W. Mansbridge.
	Apanteles cleocerdis Marsh. { (MS.)	B. A. Bower. / G. Rose.
	Microgaster minutus Reinh.	W. Mansbridge.
E. lichenea	*Ophion obscurum* Fab.	G. C. Bignell.
Miselia oxyacanthæ...	*O. luteum* L.	G. C. Bignell.
	Apanteles fulvipes Halid.	G. C. Bignell.
Phlogophora meticulosa	*Ascogaster rufidens* Wesm. ...	T. R. Billups.
	Exorista vulgaris Fallén	E. A. Fitch.

* Dipteron.
† Young larva before third moult.
(MS.) Not yet described.

Host.	Parasite.	By whom bred.
Aplecta nebulosa	*Amblyteles oratorius* Fab.......	G. C. Bignell.
	Spathius exarator L.............	R. South.
	Apanteles (sp. undescribed) ...	T. A. Chapman
Hadena adusta........	*Ichneumon fabricator* Fab. ...	J. Sang.
H. protea	*Ophion obscurum* Fab.	G. C. Bignell.
H. dentina...............	*Paniscus testaceus* Grav.	G. C. Bignell.
H. chenopodii	*Paniscus testaceus* Grav.	C. Fenn.
	Pimpla instigator Fab.	H. W. Barker.
H. oleracea	*Ichneumon fabricator* Fab. ...	C. Fenn.
	Amblyteles castigator Fab. ...	R. South.
	Exetastes osculatorius Fab. ...	G. C. Bignell.
H. pisi	*Henicospilus ramidulus* L. ...	G. C. Bignell.
	H. repentinus Holmg.	—. Lowry.
		H. Marsh.
	Ophion luteum L. .... {	T. A. Marshall.
		J. R. Wellman.
	Paniscus testaceus Grav.	H. Marsh.
	Apanteles congestus Nees	C. H. H. Walker.
	A. difficilis Nees	G. C. Bignell.
	Gonia ornata Meig.	A. Short.
H. thalassina	*Cryptus obscurus* Grav.	J. Sang.
Xylocampa lithorhiza	*Apanteles fulvipes* Halid.	G. C. Bignell.
Xylina rhizolitha......	*Paniscus testaceus* Grav.	G. C. Bignell.
	Mesochorus formosus Bridgm.	G. C. Bignell.
	Apanteles fulvipes Halid.......	G. C. Bignell.
	Mesochorus vitticollis Holmg.	G. C. Bignell.
Cucullia verbasci......	*Amblyteles palliatorius* Grav.	T. R. Billups
	Ophion longigenum Thoms. ...	B. A. Bower.
	Casinaria mesozosta Grav. ...	R. Adkin.
	†*Limneria fenestralis* Holmg.	H. W. Barker.
	Exetastes osculatorius Fab. ...	H. W. Barker.
	Rhogas circumscriptus Nees...	W.H.B.Fletcher.
	†*Apanteles ruficrus* Halid.	G. C. Bignell.
	A. fulvipes Halid.	J. N. Still.
	Microplitis vidua Ruthe	T. R. Billups.
	M. mediana Ruthe .... {	T. A. Marshall.
		J. H. Wood.
	M. tuberculifera Wesm.	G. C. Bignell.
C. gnaphalii............	*Limneria ensator* Grav.	W. H. Tugwell.
	Macrocentrus linearis Fab. ...	W. H. Tugwell.
C. chamomillæ........	*Rhogas circumscriptus* Nees ...	J. N. Still.
	Apanteles ruficrus Halid.	G. C. Bignell.
Heliothis dipsacea ...	*Anomalon cerinops* Grav.	R. Adkin.
	Schizopyga circulator Pz.	R. Adkin.
Anarta myrtilli	*Limneria ruficincta* Grav. {	G. C. Bignell.
		H. W. Barker.
	Meteorus deceptor Wesm. ...	Mrs. Hutchinson.
	M. pulchricornis Wesm.	R. W. Bowyer.

* Dipteron.
† From young larvæ four to six lines in length.
‡ Hyperparasite on *Apanteles fulvipes*.

Host.	Parasite.	By whom bred.
Brephos parthenias ...	Rhogas bicolor Spin.	C. Fenn.
	Meteorus versicolor Wesm. ...	C. Fenn.
B. notha..................	Limneria geniculata Grav. ...	R. Adkin.
Plusia orichalcea ...	Sagaritis punctata Bridgm. ...	W. Cross.
Pl. chrysitis	Hemiteles fulvipes Grav.	V. R. Perkins.
	Apanteles pallidipes Reinh. ...	G. C. Bignell.
	Meteorus versicolor Wesm. ..	J. H. Carpenter.
Pl. festucæ	Pimpla graminellæ Schrank.	W. J. Cross.
	*Nemoræa notabilis Meig.	J. R. Wellman.
Pl. iota	Tryphon elongator Fab.	R. South.
	Apanteles pallidipes Reinh. ...	G. C. Bignell.
Pl. gamma	Agrypon tenuicorne Grav. ...	H. W. Barker.
	Sagaritis zonata Grav.	T. R. Billups.
	Apanteles pallidipes Reinh. {	G. C. Bignell. E. A. Fitch. J. E. Fletcher. W.H.B.Fletcher.
	*Exorista vulgaris Fallén	W.H.B.Fletcher.
Pl. interrogationis ...	Limneria crassicornis Grav. ...	W. Fletcher.
	Apanteles fulvipes Halid.	W.F. de V. Kane.
Amphipyra pyramidea	Apanteles fulvipes Halid.	G. C. Bignell.
	A. triangulator Wesm. ...... {	J. Hellins. T. A. Chapman.
Mania typica	Ichneumon saturatorius L. ...	W. Mansbridge.
	I. lepidus Grav.	W. Mansbridge.
	Amblyteles litigiosus Grav. ...	W. Mansbridge.
	Sagaritis laticollis Holmg. ...	G. C. Bignell.
	*Exorista vulgaris Fallén	W. Mansbridge.
Toxocampa craccæ ...	Henicospilus ramidulus L. ...	R. South.
	Ophion luteum L.	R. South.
	Phytodiaëtus segmentator Grav.	R. South.
Catocala nupta	Apanteles fulvipes Halid.	G. C. Bignell.

* Dipteron.

INDEX.

PLATE LXXXVII.

EREMOBIA OCHROLEUCA.

1, 1 *a*, larvæ after last moult; on panicles of *Dactylis glomerata* in chalk-pits, July 2nd; imago emerged July 30th, 1870; also on *Avena strigosa* and *A. fatua*, July 2nd, 1875. See pp. 1—3.

DIANTHÆCIA CARPOPHAGA.

2, 2 *a*, 2 *b*, larvæ after last moult; on *Silene maritima*, July 14th, 1864; others (two) on same plant, September 15th, 1862; imagos emerging June 30th and July 12th, 1863.

DIANTHÆCIA CÆSIA.

3, 3 *a*, 3 *b*, larvæ after last moult; 3, on *Silene maritima*, July 31st, 1869; 3 *a*, on the same plant, August 9th, 1869; imagos emerged July 16th and 31st, 1869; 3 *b*, on the leaves and flowers but chiefly on the seeds of *Silene inflata*, from the Isle of Man, July 15th, 1867. See pp. 6—8.

DIANTHÆCIA IRREGULARIS.

4, 4 *a*, 4 *b*, 4 *c*, larvæ in various stages of growth; on flowers and seeds of *Silene otites*, August 20th, 1869, August 5th and 15th, 1870. See pp. 13, 14.

DIANTHÆCIA BARRETTII = LUPERINA LUTEAGO *Gn.*

5, larva after last moult; in root of *Silene maritima*, September 13th, 1878, pupated September 17th; ♀ moth bred June 27th, 1879. See pp. 9—13.

DIANTHÆCIA CAPSINCOLA.

6, 6 *a*, 6 *b*, larvæ after last moult; on seeds of *Silene inflata* and of *Lychnis diurna*, October 7th, 1862; on *Lychnis diurna*, July 18th, 1864; August 5th, 1864, imagos emerging June 14th, 1865; also July 4th, 1859, and September 2nd to 12th, 1863.

DIANTHÆCIA CUCUBALI.

7, 7 *a*, 7 *b*, larvæ after last moult; on leaves of *Silene inflata*, July 11th, 1859, July 28th, 1861, and July 12th to 18th, 1864.

PLATE LXXXVIII.

HECATERA DYSODEA.

1, 1 *a*, 1 *b*, 1 *c*, larvæ after last moult; on flowers and seeds of lettuce, August 31st, 1866.

HECATERA SERENA.

2, 2 *a*, 2 *b*, 2 *c*, 2 *d*, 2 *e*, larvæ after last moult; 2, on flowers of *Crepis taraxifolia*, August 30th, 1872; 2 *e*, August 7th, 1865, imago emerging July 10th, 1866; 2 *a*, August 5th, 1865, imago July 8th, 1866; one or more of the others on flowers and seeds of *Crepis taraxifolia*, August 1st, 1864, and on flowers and seeds of lettuce, August 19th, 1864, imago emerging July 1st, 1865.

POLIA CHI.

3, 3 *a*, 3 *b*, larvæ after last moult; on willow, June 10th, 1861, imago August 10th, 1861; on hawthorn and sallow, May 10th, 1862; and on sallow, May 27th, 1870. See pp. 15—17.

POLIA FLAVOCINCTA.

4, young larva; 4 *a*, 4 *b*, larvæ after last moult; 4 and 4 *b*, on rest-harrow, June 18th and 25th, 1863; 4 *a*, on grass, wild mint, scabious, apricot leaves, sweet-peas, July 4th, 1868. See pp. 17, 18.

POLIA NIGROCINCTA.

5, 5 *a*, 5 *b*, larvæ in various stages of growth; on flowers of *Statice armeria* and of *Silene maritima*, Isle of Man, July 12th, 1869, and eating unripe seeds of *Plantago maritima*, Douglas, Isle of Man, July 13th and 14th, 1876. See pp. 18, 19.

Plate LXXXVIII

1

1a

1c

1b

2

2c

2b

2e

2d

2a

3

3a

3b

4a

4

4b

5b

5

5a

F.C.Moore lith

W. BUCKLER del

West,Newman imp

PLATE LXXXIX.

Dasypolia templi.

1, 2, 3, 4, larvæ in various stages of growth ; 5, pupa ; figured July 11th, in the stem of *Heracleum sphondylium ;* July 20th in the root-crown ; July 26th in the hole shown in the plate.   See pp. 19—24.

Plate LXXXIX

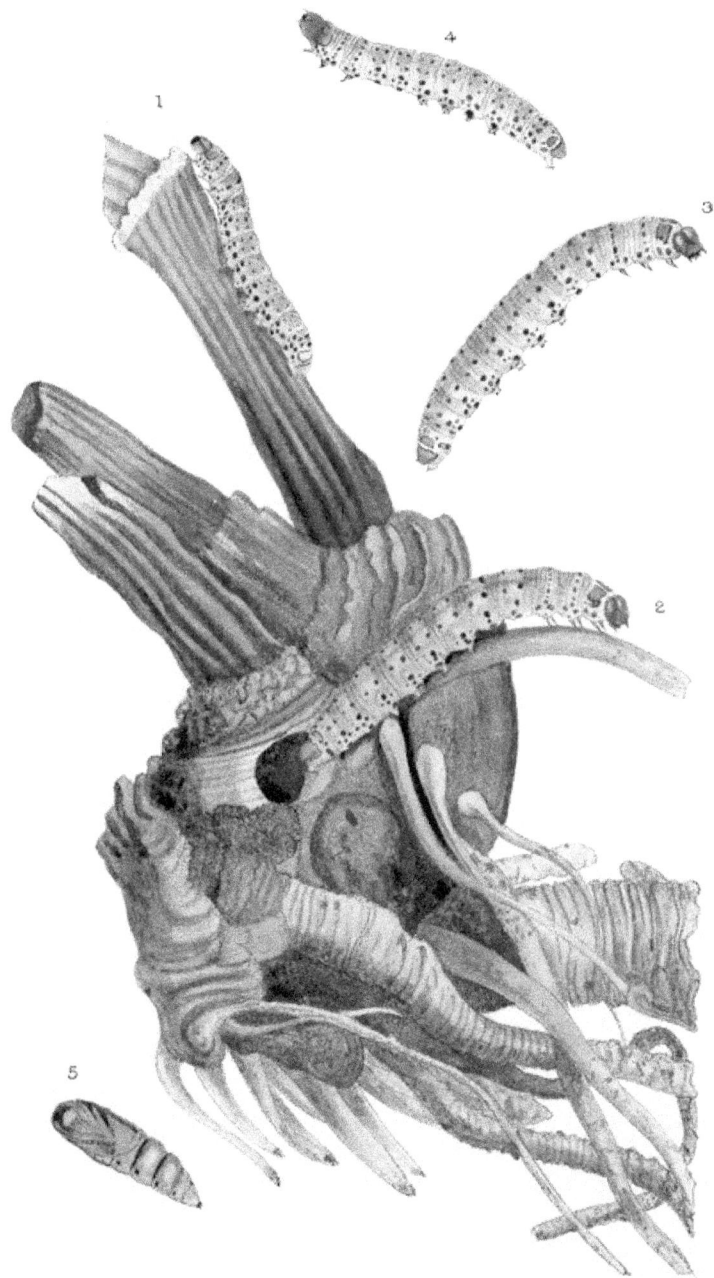

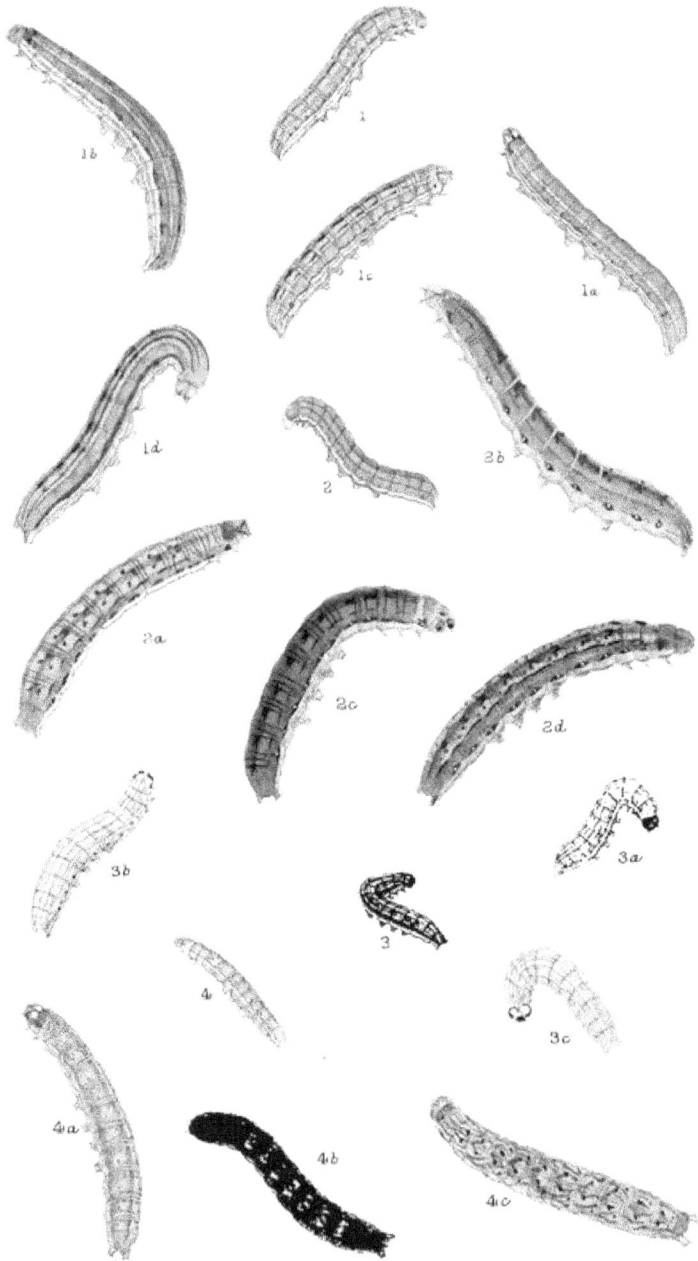

Plate XC.

1b

1

1c

1a

1d

2

2b

2a

2c

2d

3b

3a

3

4

3c

4a

4b

4c

F C Moore lith.

W. BUCKLER del.

West, Newman imp.

PLATE XC.

EPUNDA LUTULENTA.

1, 1 *a*, 1 *b*, 1 *c*, 1 *d*, larvæ in various stages ; 1 *c*, on heather and grass, June 19th, 1869 ; 1 *d*, on plantain, April 22nd, 1874 ; the others on grass, February 20th, May 8th and 20th, and June 1st, 1869. See pp. 24—27.

EPUNDA NIGRA.

2, 2 *a*, 2 *b*, 2 *c*, 2 *d*, larvæ in various stages ; 2, April 30th, when grown became like 2 *b* ; 2 *b*, on hawthorn, May 12th, 1866 ; 2 *a*, from Torquay, May 5, imago October 14th, 1856 ; 2 *c*, on hawthorn and *Galium*, April 28th, imago, October 28th, 1866 ; 2 *d*, on haw-thorn, May 5th, 1866 ; other examples drawn but not reproduced here, on sloe, would eat sallow, hawthorn, and grass, April 22nd, 1866. See pp. 27—30.

EPUNDA VIMINALIS.

3, 3 *a*, 3 *b*, 3 *c*, larvæ in various stages ; 3, 3 *a*, 3 *c*, on sallow, May 21st, imago July 14th, 1867 ; 3 *b*, on long-leaved sallow, June 3rd, imago July 13th, 1864; specimens not figured, on withy, June 27th, 1860, and on sallow, June 6th, imago July 10th, 1863.

EPUNDA LICHENEA.

4, 4 *a*, 4 *b*, 4 *c*, larvæ in various stages ; 4 *b*, on haw-thorn, April 21st, imago October 27th, 1866 ; the others on groundsel and dock, etc., April 5th, 1860 ; an example not figured, on groundsel, April 6th, imago emerging October 16th, 1868.

PLATE XCI.

Miselia oxyacanthæ.

1, 1 *a*, 1 *b*, 1 *c*, larvæ in various stages ; 1, June 7th, 1859, imago emerging October 5th, 1859 ; 1 *a*, figured June 6th, 1860 ; 1 *b*, June 8th, 1866 ; 1 *c*, June 7th, 1860.

Agriopis aprilina.

2, 2 *a*, larvæ after last moult; found in crevice of bark on oak, May 27th, imago emerging September 25th, 1861 ; and on oak June 15th, 1861. See p. 30.

Phlogophora meticulosa.

3, 3 *a*, 3 *b*, 3 *c*, 3 *d*, 3 *e*, 3 *f*, 3 *g*, larvæ in various stages ; 3, on blackthorn ; 3 *a*, figured March 1st, 1862 ; 3 *f*, on flowers of chrysanthemum, January 17th, 1868 ; 3 *g*, variety, August 5th, imago September 25th, 1859 ; two of the others on dock, July 11th, imago emerging August 11th, 1864 ; and on geranium leaves, November 5th, 1867. See pp. 30—33.

Phlogophora empyrea.

4, 4 *a*, larvæ after last moult; reared from the egg on *Ranunculus repens* and *R. ficaria*, March 21st to 24th, 1874.

Euplexia lucipara.

5, 5 *a*, larvæ after last moult ; 5 *a*, on fern and sallow, September 12th, 1863 ; 5, on fern, August 17th, 1878 ; an example not figured on sallow, August 24th, 1861, imago emerging June 1st, 1862.

Plate XCI

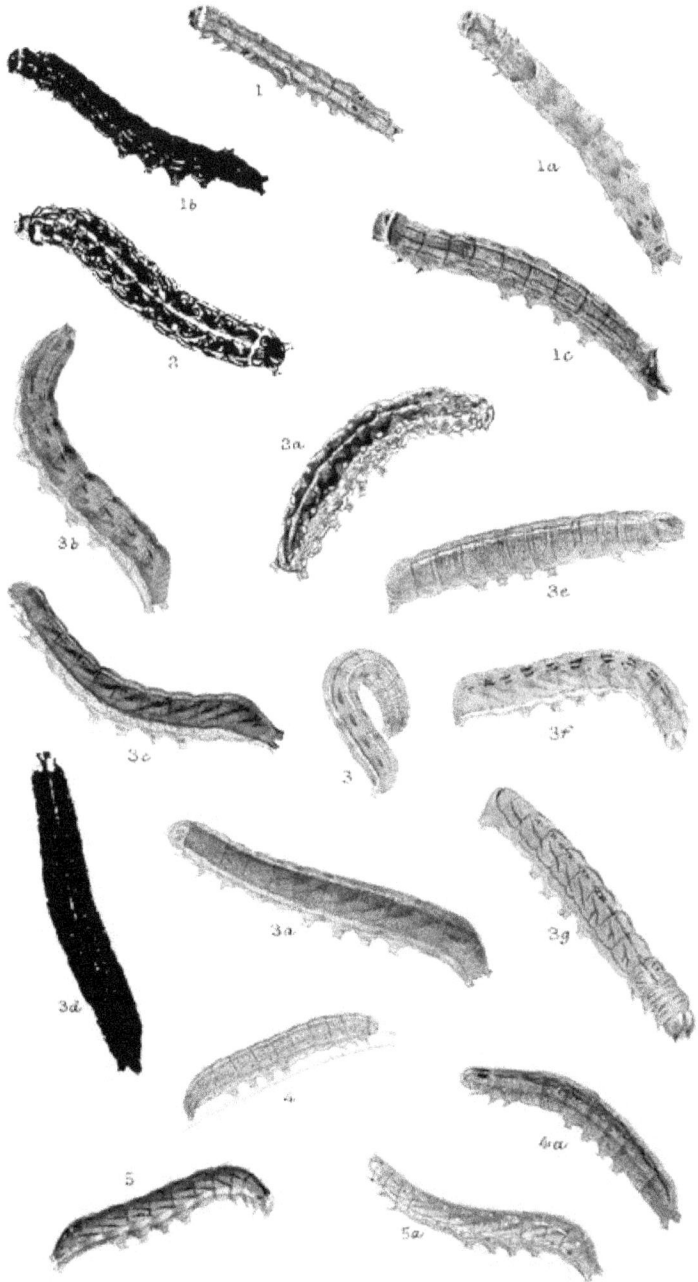

Plate XCII

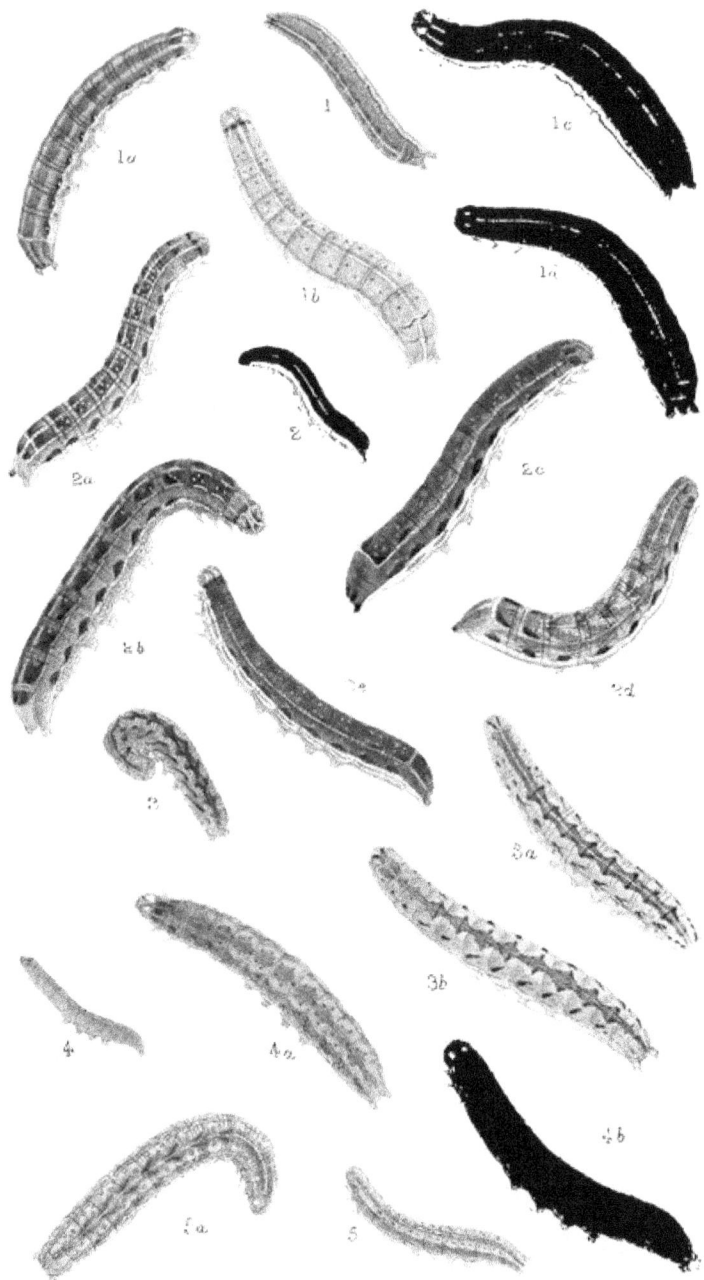

PLATE XCII.

Aplecta herbida.

1, 1 *a*, 1 *b*, 1 *c*, 1 *d*, larvæ in various stages of growth; October to April, on dock, nettle, chickweed, sallow buds, and grass; March 12th, imago May 24th, 1862; 1 *a*, on *Plantago major*, reared from a quarter of an inch long, figured 25th November, 1875, pupa figured 5th February, 1876; imago appeared June 19th, 1876. See pp. 36—39.

Aplecta occulta.

2, 2 *a*, 2 *b*, 2 *c*, 2 *d*, 2 *e*, larvæ in various stages of growth; from eggs reared on knot-grass and dock, October 16th, 27th, November 1st, 1869, February 28th, March 1st, 1870; 2, small one figured before hybernating, from heather, Ross-shire, 6th October, 1868; 2 *c* and 2 *e*, on heather and dock, sallow and bramble, May 1st and 5th, 1869; another figured October 7th and 10th; another figured as pupa on September 30th, 1874; imagos emerged October 13th to December 22nd, 1874. See pp. 39—43.

Aplecta nebulosa.

3, 3 *a*, 3 *b*, larvæ in various stages; on thorn and bramble, 1861; on sallow, April 1st and 20th, imago appearing June 9th; on wild cherry, April 28th, 1864; 3, on hawthorn buds, March 28th, 1868.

Aplecta tincta.

4, 4 *a*, 4 *b*, larvæ in various stages; 4, one reared from egg, figured in third moult, September 10th, 1874, fed on knot-grass and birch; 4 *a*, 4 *b*, on birch and dock, April 26th; on hawthorn, May 5th, imagos appearing June 19th and 21st, 1862. See pp. 44, 45.

Aplecta advena.

5, 5 *a*, larvæ in various stages; on knot-grass, July 26th and September 1st, 1865. See pp. 45—47.

PLATE XCIII.

HADENA ADUSTA.

1, 1 *a*, 1 *b*, 1 *c*, 1 *d*, larvæ in various stages; reared on lettuce, October 19th, 1860; reared from the egg on lettuce, and figured August 22nd, 1863; 1 *a*, on knot-grass, sallow, and hawthorn, September 15th, 1865; 1 *b*, September 11th, 1869; 1 *d*, on *Silene inflata*, October 11th, 1878. See pp. 47, 48.

HADENA PROTEA.

2, 2 *a*, larvæ after last moult; 2, on oak, May 18th, 1867, imago September 5th, 1867; 2 *a*, reared from the egg on oak, June 2nd, imago September 7th, 1883. See pp. 48, 49.

HADENA GLAUCA.

3, 3 *a*, larvæ in various stages; 3, on tops of weeping willow, July 30th, 1860; 3 *a*, found and reared on sallow, figured July 18th, 1860, imago appearing June, 1861.

HADENA DENTINA.

4, 4 *a*, 4 *b*, 4 *c*, larvæ in various stages, figured from August 3rd to 19th, 1863; reared on flower-heads of dandelion and plantain; imagos appeared July 10th to 14th, 1864. See pp. 50—52.

HADENA CHENOPODII.

5, 5 *a*, 5 *b*, 5 *c*, 5 *d*, larvæ after last moult; figured September 16th, 1860, imagos appearing July 15th and 19th, 1861; and figured on August 17th and 24th, 1861, from larvæ fed on *Chenopodium botryoides* and *Atriplex patula*.

F.C.Moore lith.

W.S. Plowman imp

Plate XCIV.

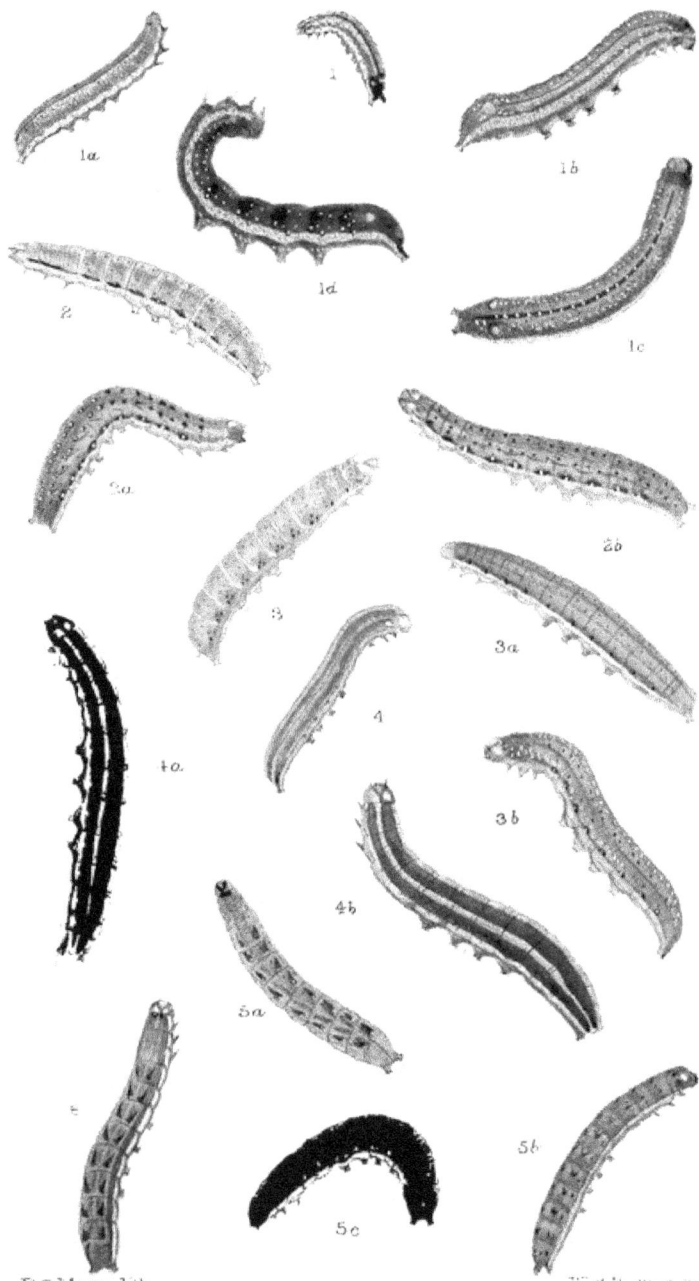

1a
1
1b
1d
1c
2
2a
2b
3
3a
4a
4
3b
4b
5a
5
5b
5c

F.C.Moore lith.

W.BUCKLER. del.

PLATE XCIV.

HADENA ATRIPLICIS.

1, 1 *a*, 1 *b*, 1 *c*, 1 *d*, larvæ in various stages of growth ; on knot-grass and persicaria ; figured July 31st, August 4th, 6th, 8th, and 16th, 1873 ; imagos emerged June 16th and 20th, 1874.

HADENA SUASA.

2, 2 *a*, 2 *b*, larvæ after last moult ; on *Plantago major* and knot-grass ; figured July 18th and 23rd, 1866 ; one imago appeared August 23rd, 1866, another May 21st. See pp. 52, 53.

HADENA OLERACEA.

3, 3 *a*, 3 *b*, larvæ after last moult ; on groundsel, chickweed, etc. ; figured August 19th, 1859 ; also on goosefoot and dock, figured September 18th, 1872, imago emerging July 3rd, 1873.

HADENA PISI.

4, 4 *a*, 4 *b*, larvæ in various stages ; on young shoots of weeping willow ; figured July 31st, 1860 ; 4, on *Pteris aquilina*, figured September 9th, 1873.

HADENA THALASSINA.

5, 5 *a*, 5 *b*, 5 *c*, larvæ after last moult ; several on knot-grass, figured July 26th, 1865, imagos emerging May 31st to June 4th, 1866 ; 5, on hawthorn, figured August 15th, 1861, imago emerging June 14th, 1862 ; 5 *b*, one of two on apple, oak, etc., from Mrs. Hutchinson, figured September 2nd, 1879, bred June 1st, 1880. See pp. 53—55.

PLATE XCV.

HADENA CONTIGUA.

1, 1 *a*, 1 *b*, larvæ after last moult; on nut, dock, etc. ; figured September 16th, 1860, and September 16th, 1863 ; imago emerged June 17th, 1861.

HADENA GENISTÆ.

2, 2 *a*, 2 *b*, 2 *c*, 2 *d*, larvæ in various stages of growth ; on chickweed and persicaria ; figured July 14th and 29th to August 7th, 1865 ; imago emerged June 11th to 13th, 1866. See pp. 55, 56.

HADENA RECTILINEA.

3, 3 *a*, 3 *b*, 3 *c*, larvæ in various stages of growth ; on sallow ; full-grown September 23rd, 1864, hybernating till the spring, but not feeding again ; a young larva found on balsam (*Impatiens*), July 15th, 1876, fed up on whortleberry in autumn ; imago bred June, 1877. See pp. 56—58.

XYLOCAMPA LITHORHIZA.

4, 4 *a*, 4 *b*, larvæ after last moult; on woodbine, June 14th, 1860, imago emerged March 4th, 1861 ; on May 28th, 1864 ; 4 *b*, on honeysuckle, June 24th, 1867.

CLOANTHA SOLIDAGINIS.

5, 5 *a*, larvæ after last moult ; on whortleberry and cowberry, July 5th, 1862, and on bilberry, May 20th, 1866. See pp. 58, 59.

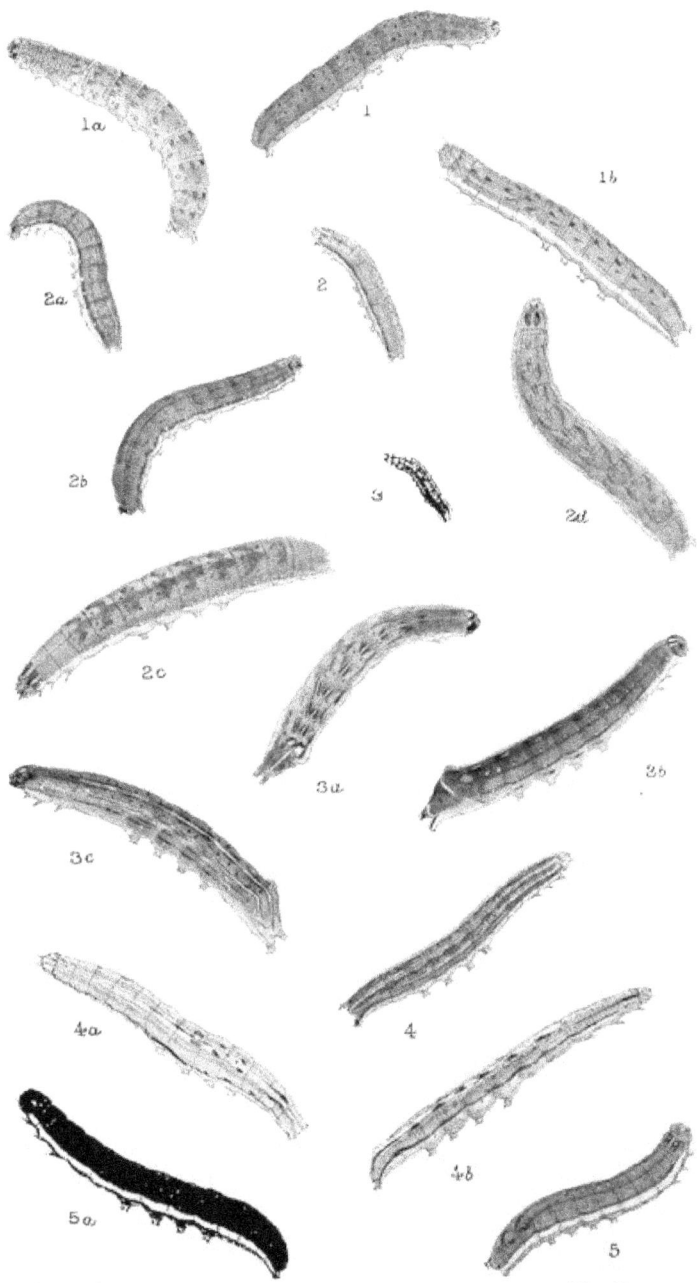

1a   1   1b
2a   2   2d
2b   3
2c   3a   3b
3c
4a   4
5a   4b   5

F.C.Moore lith.   W.BUCKLER del   West Newman imp

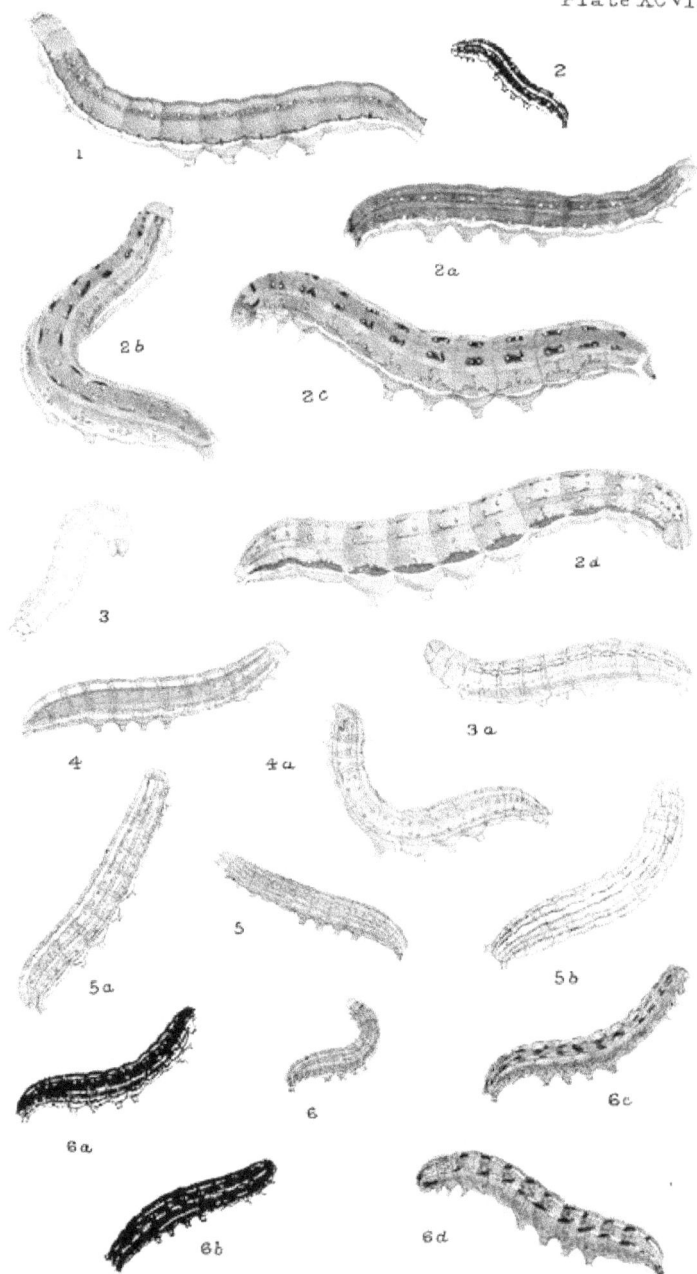

Plate XCVI.

1

2

2a

2b

2c

2d

3

3a

4

4a

5

5a

5b

6

6a

6b

6c

6d

F.C.Moore lith.

W.BUCKLER del

West, Newman imp.

PLATE XCVI.

CALOCAMPA VETUSTA.

1, larva after last moult; on persicaria, sorrel, dock, etc.; July 13th, 1868.

CALOCAMPA EXOLETA.

2, 2 *a*, 2 *b*, 2 *c*, 2 *d*, larvæ in various stages of growth; on knot-grass and dock, June 3rd, 11th, and 27th, 1868; and young larva on *Stachys sylvatica* and bryony, June 20th to 27th, 1874.

XYLINA RHIZOLITHA.

3, 3 *a*, larvæ after last moult; a cannibal; on oak, June 7th, 1861, June 23rd, 1863, and June 7th, 1871; imagos emerged October 5th, 1863, and October 16th, 1871. See pp. 60—62.

XYLINA SEMIBRUNNEA.

4, 4 *a*, larvæ after last moult; reared from eggs on *Fraxinus excelsior*; figured June 16th, 1870; imagos emerged September 21st to 27th, 1870. See pp. 62, 63.

XYLINA PETRIFICATA.

5, 5 *a*, 5 *b*, larvæ in various stages of growth; a cannibal; on sallow, June 21st, 1861; and two on ash, June 8th and 13th, 1867; imago emerged November 10th, 1867.

XYLINA CONFORMIS.

6, 6 *a*, 6 *b*, 6 *c*, 6 *d*, larvæ in various stages of growth; on alder, May 27th, June 2nd to 7th and 17th, 1871, and two larvæ June 28th, 29th, 1872; imago emerged August 17th, 1871. See pp. 63—67.

PLATE XCVII.

CUCULLIA VERBASCI.

1, 1 *a*, 1 *b*, 1 *c*, 1 *d*, larvæ in various stages of growth; on *Scrophularia nodosa, S. aquatica* and *Verbascum thapsus*, July 12th, 13th, and 22nd, 1867, and on *S. aquatica*, July 2nd, 1867; 1 *a*, on *Verbascum lychnitis*, June 23rd, 1874; imago emerged May 15th, 1875; 1 *d*, on *V. thapsus*, June 21st, 1858; imago emerged April 28th, 1859. See p. 67.

CUCULLIA SCROPHULARIÆ.

2, 2 *a*, 2 *b*, 2 *c*, larvæ after last moult; on *Scrophularia nodosa*, July 4th and 8th, 1867. See pp. 68, 69.

CUCULLIA LYCHNITIS.

3, 3 *a*, 3 *b*, 3 *c*, 3 *d*, larvæ after last moult; on *Verbascum nigrum*, September 5th, 1861, August 5th, 7th, and 21st, 1863; imago emerged June 9th, 1865.

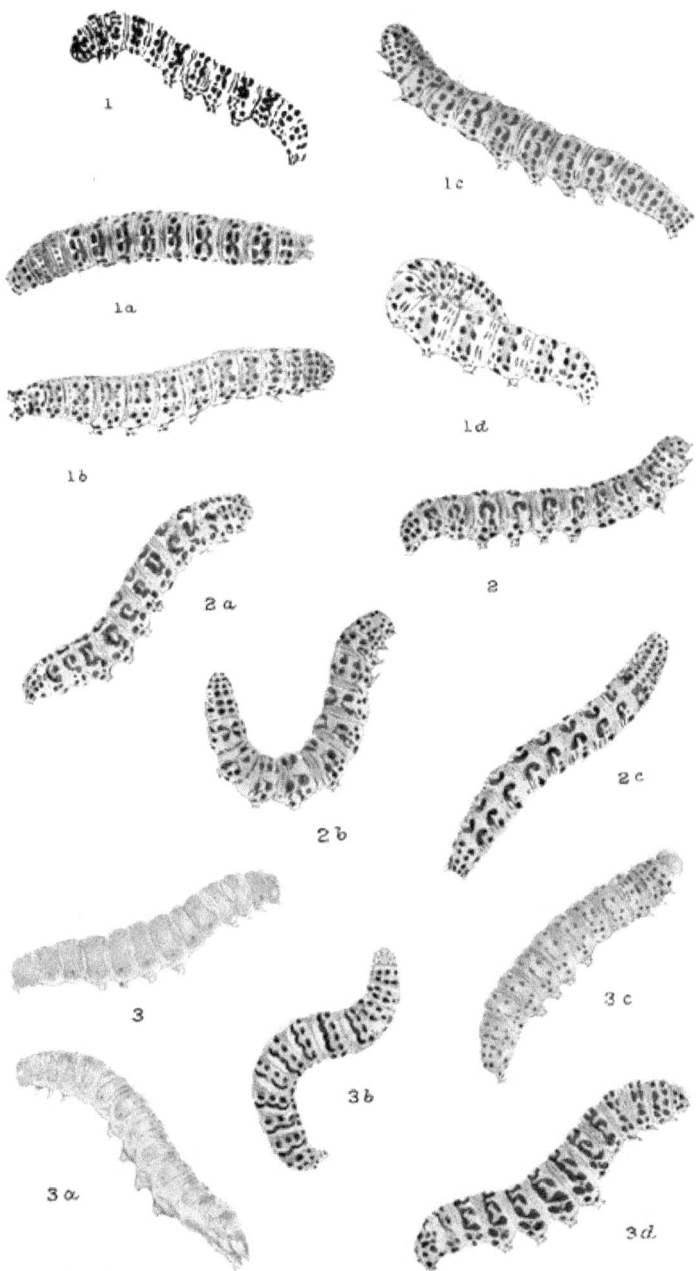

1

1c

1a

1d

1b

2a

2

2b

2c

3

3b

3c

3a

3d

F.C. Moore lith.

West, Newman imp.

W.BUCKLER del.

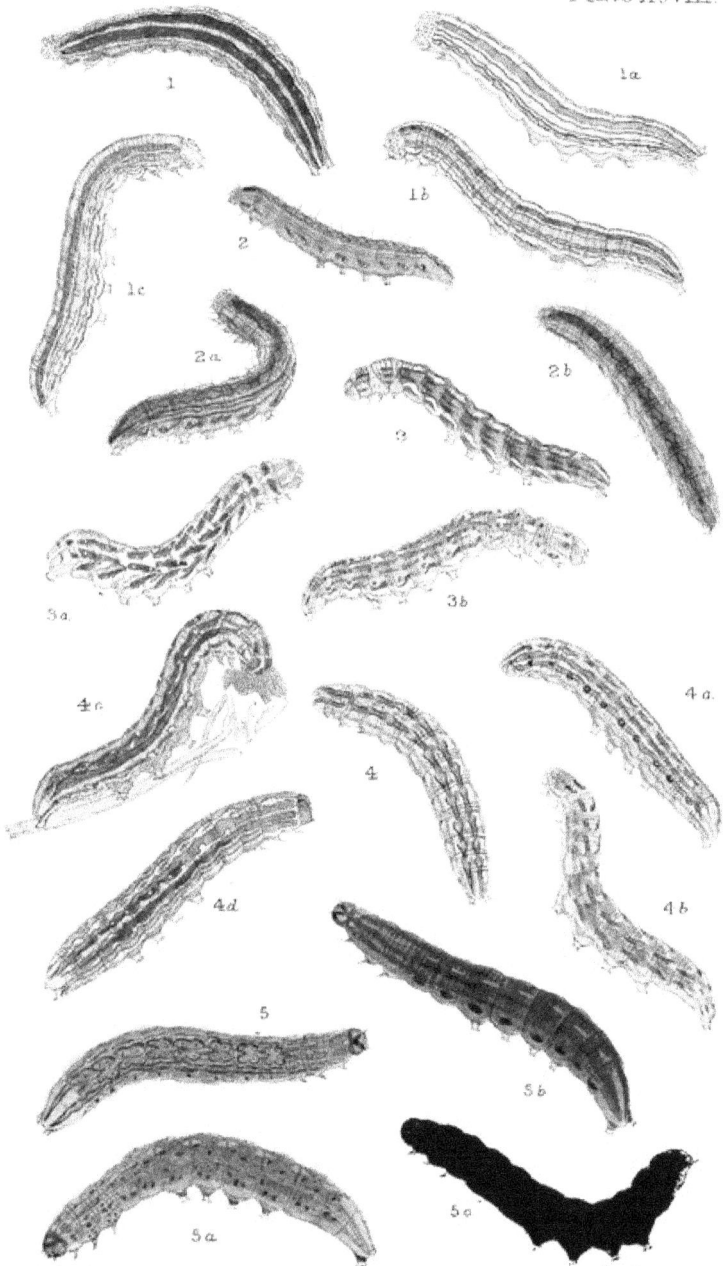

Plate XCVIII.

1

1a

1b

2

1c

2a

2b

3

3a

3b

4c

4a

4

4d

4b

5

5b

5a

5a

F C Moore lith.

West, Newman imp.

W. BUCKLER del.

PLATE XCVIII.

CUCULLIA ASTERIS.

1, 1 *a*, 1 *b*, 1 *c*, larvæ after last moult; on flowers of golden-rod, September 20th, 1867 ; 1 *a*, on China aster, eating the flower ; 1 *c*, on golden-rod, August 22nd, 1871 ; imago emerged July 24th, 1872.

CUCULLIA GNAPHALII.

2, 2 *a*, 2 *b*, larvæ after last moult ; on leaves of *Solidago virgaurea*, September 26th, 1869, August 26th and September 2nd, 1871. See pp. 69—71.

CUCULLIA ABSINTHII.

3, 3 *a*, 3 *b*, larvæ after last moult; on seeds of wormwood, September 13th, 1862; on flowers and seeds of wormwood, September 3rd and 8th, 1866 ; imago emerged July 29th, 1867.

CUCULLIA CHAMOMILLÆ.

4, 4 *a*, 4 *b*, 4 *c*, 4 *d*, larvæ after last moult ; on wild chamomile flowers ; Plymouth, June 24th to July 6th, 1863 ; Torquay, July 9th, 1870, imago May 21st; on *Anthemis cotula*, July 23rd, 1870.

CUCULLIA UMBRATICA.

5, 5 *a*, 5 *b*, 5 *c*, larvæ after last moult ; on sow-thistle, hiding from daylight, August 18th to September 1st, 1866 ; two on *Lactuca virosa*, August 5th to 12th, 1874 ; ♂ ♀ imagos emerged, July 13th, 1875 ; and one larva August 15th, 1864. See pp. 71—73.

PLATE XCIX.

HELIOTHIS MARGINATA.

1, 1 *a*, 1 *b*, 1 *c*, 1 *d*, 1 *e*, larvæ in various stages of growth ; on thorny rest-harrow, August 24th, 1861 ; on birch, August 3rd, 1861, ♀ imago emerging June 25th, 1863 ; on *Ononis arvensis*, August 31st, 1861 ; on nut ; on rest-harrow, but fed on roses, birch, and nut, and fond of rest-harrow flowers, September 19th, 1863, imago emerging June 18th, 1864 ; on August 15th, 1867 ; 1, on sallow, August 25th, 1871, imago emerging July 17th, 1872.

HELIOTHIS PELTIGERA.

2, 2 *a*, 2 *b*, 2 *c*, 2 *d*, 2 *e*, larvæ after last moult; on henbane and flowers of rest-harrow, July 29th, 1862 ; on henbane, July 13th, 16th, 20th, and 22nd, imagos emerging September 9th and 14th, 1868; one on *Datura stramonium*, October 16th, 1867.

HELIOTHIS DIPSACEA.

3, 3 *a*, 3 *b*, 3 *c*, 3 *d*, 3 *e*, larvæ in various stages of growth ; 3, on purple-clover flower, August 27th, 1867 ; 3 *a*, on rest-harrow, August 28th, 1873, littoral, Slapton-on-Sea ; 3 *b*, on *Crepis virens*, September 11th, 1873, imago out July 10th, 1874 ; 3 *c*, on flowers and seeds of *Ononis arvensis*, September 12th, 1873 ; 3 *d*, on sorrel and toadflax seeds, September 15th, 1870 ; 3 *e*, on seed-capsules of *Silene otites*, August 25th, 1870. See pp. 75—78.

Plate XCIX.

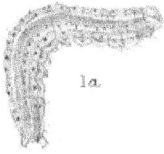

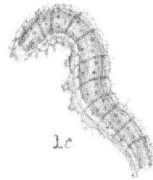

1a.

1

1c

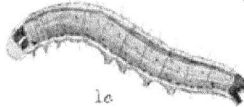

1b

1e

1d

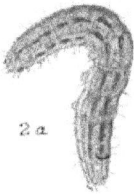

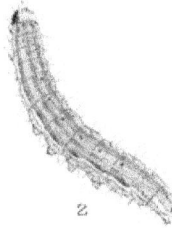

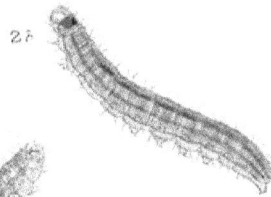

2a

2

2b

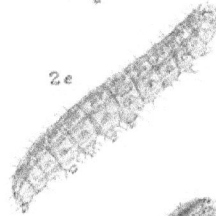

2e

2c

2d

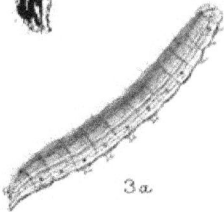

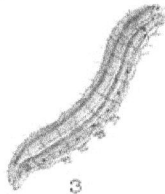

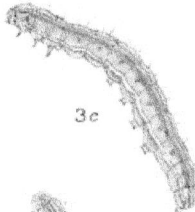

3a

3

3c

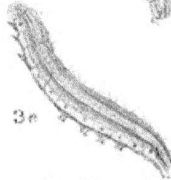

3b

3d

3e

F.C Moore lith.

West, Newman imp.

W. BUCKLER. del.

PLATE C.

ANARTA MELANOPA.

1, 1 *a*, 1 *b*, larvæ in various stages of growth; reared from eggs on sallow, July 4th, 8th, and 14th, 1875. See pp. 78—80.

ANARTA CORDIGERA.

2, 2 *a*, 2 *b*, larvæ in various stages of growth; on *Arbutus uva-ursi*, fed well on *A. unedo*, July 4th, 9th, 16th, and 17th, 1875; Rev. John Hellins bred the moth on May 1st, 1876, having raised these larvæ from eggs laid on the 1st of June, 1875. See pp. 80—82.

ANARTA MYRTILLI.

3, 3 *a*, larvæ after last moult; on ling, September 15th, 1869.

HELIODES ARBUTI.

4, 4 *a*, 4 *b*, 4 *c*, larvæ in various stages of growth; feeding in seed-capsules of *Cerastium vulgatum*, from eggs laid in the flowers; figured June 11th, 13th, 15th, and 18th, 1881. See pp. 83—88.

AGROPHILA SULPHURALIS.

5, 5 *a*, 5 *b*, 5 *c*, larvæ in various stages of growth; on *Convolvulus arvensis*, July 26th and August 7th, 1867, August 19th, 1871, and July 15th, 1872. See pp. 89, 90.

PLATE CI.

ACONTIA LUCTUOSA.

1, 1 *a*, larvæ after last moult; on *Convolvulus arvensis*, July 8th, 1868. See pp. 90—92.

ERASTRIA VENUSTULA.

2, 2 *a*, larvæ after last moult; on flowers of *Tormentilla officinalis*, July 22nd, 1862; imago emerged June 5th, 1863.

ERASTRIA FUSCULA.

3, 3 *a*, 3 *b*, larvæ in various stages of growth; on *Molinia cærulea*, September 15th, 18th, 22nd, and 26th, 1873; imago emerged June 2nd, 1874. See pp. 29, 93.

BANKIA BANKIANA.

4, 4 *a*, larvæ in various stages of growth; from eggs on *Poa annua*, sent from Ely by Rev. G. H. Raynor, July 18th and 31st, 1882; one male bred 29th June, 1883. See pp. 94—96.

HYDRELIA UNCANA.

5, 5 *a*, larvæ in various stages of growth; on *Carex sylvatica*, July 28th, 1868; full-grown, August 8th, 1868. See pp. 96, 97.

BREPHOS PARTHENIAS.

6, 6 *a*, 6 *b*, larvæ after last moult; on birch, June 13th, 1861, and June 16th, 1862; imagos emerged March 26th, 1862, and June 15th, 1863.

BREPHOS NOTHA.

7, 7 *a*, 7 *b*, 7 *c*, 7 *d*, larvæ in various stages of growth; on aspen, June 2nd and 12th, 1869, June 7th and 12th, 1869, June 10th, 1870; imagos emerged April 9th, 1870, and April 4th and 7th, 1871. See pp. 98—100.

1

1 a.

2

2 a.

3

3 a.

3 b

4

4 a

5

5 a.

5 a.

6 a

6

6 b

7 b

7 a.

7 c

7 d

F.W.Frohawk lith

W. BUCKLER del.

West, Newman imp.

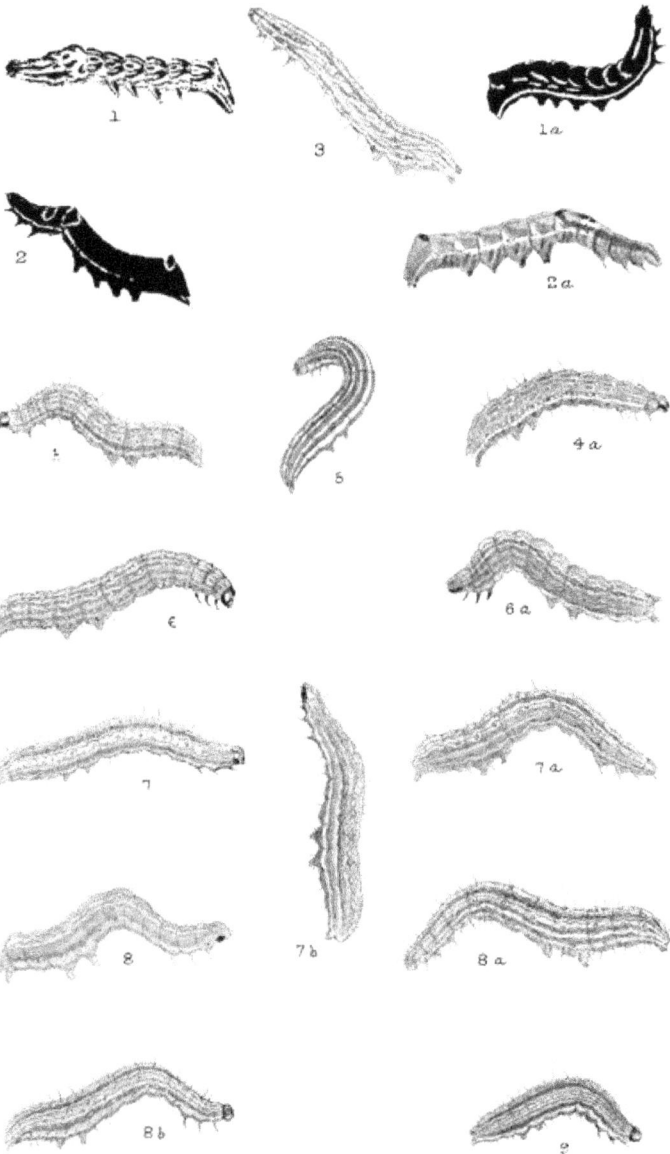

Plate CII.

1
3
1a
2
2a
4
5
4a
6
6a
7
7a
7b
8
8a
8b
9

F.W.Frohawk lith.

West, Newman imp.

W. BUCKLER del.

PLATE CII.

ABROSTOLA URTICÆ.

1, 1 *a*, larvæ after last moult; on stinging-nettle, August 30th, 1860, and July 31st, 1861; imago June 1st, 1862.

ABROSTOLA TRIPLASIA.

2, 2 *a*, larvæ after last moult; on hop and nettle, August 18th, 1862; imagos, June 18th to 30th, 1863.

PLUSIA CHRYSITIS.

3, larva after last moult; on stinging-nettle, May 23rd, 1874; another on stinging-nettle, April 15th, imago June 5th, 1869.

PLUSIA BRACTEA.

4, 4 *a*, larvæ after last moult; on *Lamium album* and stinging-nettle, April 30th and May 6th; imago June 20th, 1873. See pp. 103—107.

PLUSIA FESTUCÆ.

5, larva after last moult; reared on coarse grasses and *Sparganium*, May 8th, 1866.

PLUSIA IOTA.

6, 6 *a*, larvæ after last moult; 6, reared from eggs, May 17th, 1875; 6 *a*, on white dead-nettle, February 19th, 1862; others on *Lamium album* and *L. purpureum*, honeysuckle, and cow-parsley, April 22nd, imago June 4th, 1875. See pp. 107—110.

PLUSIA PULCHRINA.

7, 7 *a*, 7 *b*, larvæ after last moult; on groundsel, April 26th, imagos June 12th to 16th, 1862; on *Geum urbanum*, April 8th, imago June 5th, 1868; on cow-parsley, *Lamium album*, and honeysuckle, May 6th, 1876. See pp. 110—112.

PLUSIA GAMMA.

8, 8 *a*, 8 *b*, larvæ after last moult; 8, August 1st, 1859; 8*a*, on clover flowers, July 13th, imago August 9th, 1864; 8 *b*, on *Convolvulus arvensis*, July 21st, imago August 10th, 1872. See pp. 112—115.

PLUSIA INTERROGATIONIS.

9, larva after last moult; on heather, June 10th, imago July 8th, 1869. See pp. 115—117.

PLATE CIII.

Gonoptera libatrix.

1, larva after last moult; on sallow, August 3rd, 1860; imago September 12th, 1860.

Amphipyra pyramidea.

2, 2 *a*, 2 *b*, larvæ after last moult; 2 *a*, on oak, June 8th, 1860; 2 *b*, on rose, June 7th, imago July 31st, 1865.

Amphipyra tragopogonis.

3, 3 *a*, 3 *b*, larvæ after last moult; 3 *a*, on sallow, June 13th, imago July 29th, 1874; 3 *b*, on garden fennel; 3 and another, June 12th—15th, imagos August 8th—15th, 1861; and June 23rd, imago August 10th, 1863; on garden strawberry.

Mania typica.

4, young larva; 4 *a*, 4 *b*, 4 *c*, larvæ after last moult; on buds of sallow, April 15th; imago July 7th, 1861; also April 29th to May 8th, 1861, on dock, grass, primrose; imagos June 26th to 30th, 1861.

Mania maura.

5, young larva; 5 *a*, 5 *b*, larvæ after last moult; on hawthorn, lettuce, and dock, April 30th to May 1st; imagos July 15th to 24th, 1862.

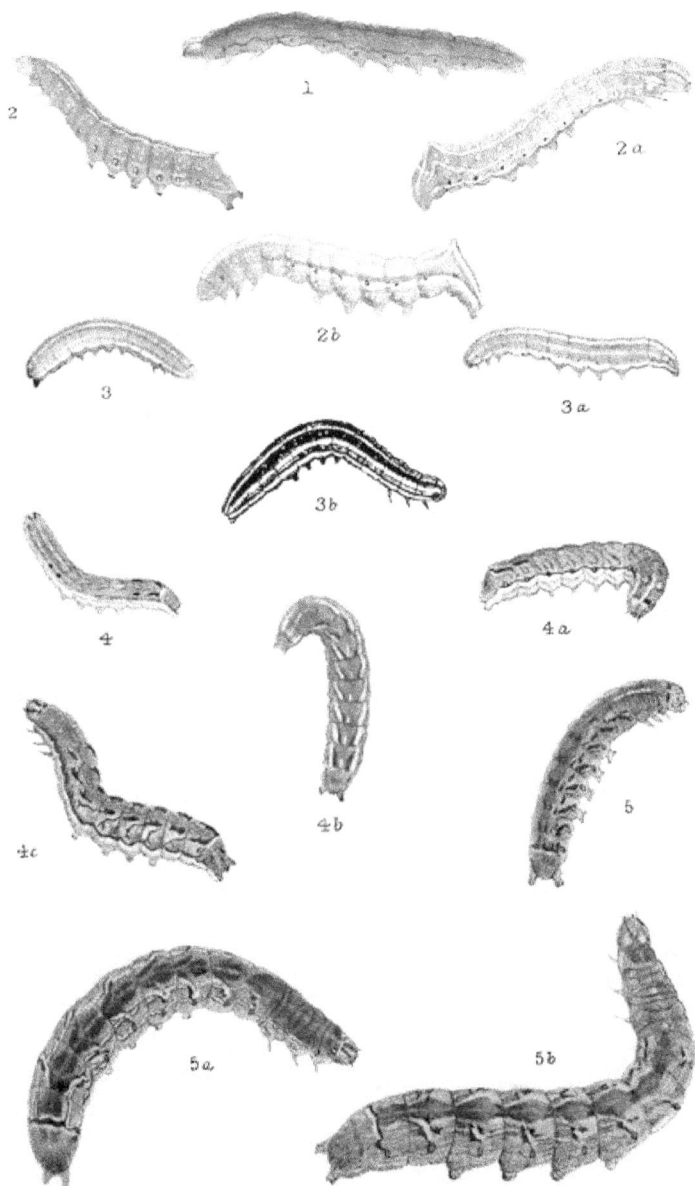

1

2

2 a

2 b

3

3 a

3 b

4

4 a

4 c

4 b

5

5 a

5 b

F.W.Frohawk lith.

West,Newman imp.

W.BUCKLER. del.

Plate CIV.

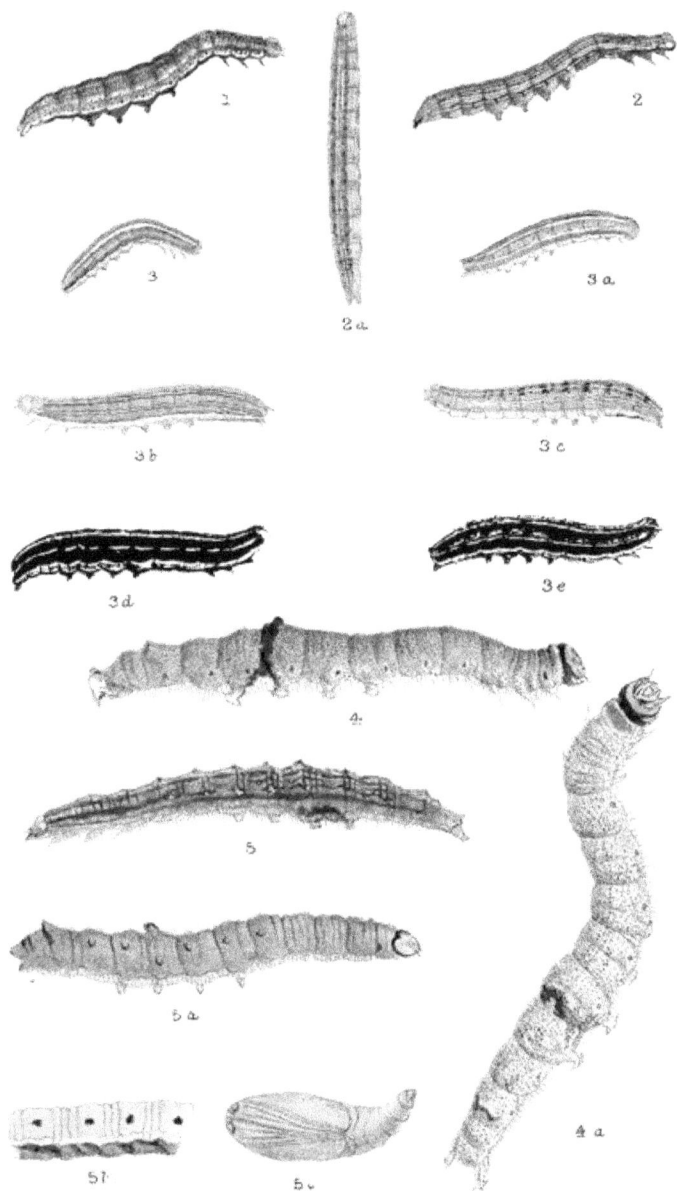

F.W.Frohawk lith.

West, Newman imp

W.BUCKLER del.

PLATE CIV.

TOXOCAMPA PASTINUM.

1, larva after last moult; on *Vicia cracca*, May 19th; imago July 7th, 1866.

TOXOCAMPA CRACCÆ.

2, 2 *a*, larvæ after last moult; near Lynmouth, Lynton, North Devon, on *Vicia sepium*, proper food *V. sylvatica*, June 20th; imago July 25th, 1865. See pp. 117, 118.

STILBIA ANOMALA.

3, 3 *a*, 3 *b*, 3 *c*, 3 *d*, 3 *e*, larvæ in various stages of growth; 3 *b*, 3 *c*, 3 *d*, 3 *e*, after last moult; January 21st, February 3rd to 8th, and February 10th; full-fed February 28th; imago September 4th, 1865. See pp. 118, 119.

CATOCALA FRAXINI.

4, larva after fifth moult, June 16th, 1881; 4 *a*, larva after last moult, June 30th, 1881; reared from German ova on *Populus fastigiata*; three moths bred, a female on August 10th, a male August 14th, and a female August 19th, 1881. See pp. 119—121.

CATOCALA NUPTA.

5, 5 *a*, larvæ after last moult; 5 *b*, flat under-side; 5 *c*, pupa; 5, on willow, June 17th, 1859; imago August 15th, 1859; 5 *a*, figured July 4th, 1870. See p. 121.

PLATE CV.

CATOCALA PROMISSA.

1, young larva ; 1 *a*, 1 *b*, larvæ after last moult ; 1 *c*, pupa ; 1, 1 *a*, on oak, from eggs, May 22nd, 24th, 27th ; imago out July 24th, 1876 ; 1 *b*, beaten from oak by Mr. F. F. Freeman, June 9th, 1882 ; male imago bred August 9th, 1882. See pp. 121—127.

CATOCALA SPONSA.

2, 2 *a*, larvæ after last moult ; 2 *b*, pupa ; 2 and another, from eggs laid in the beginning of September, 1865 ; hatched in April, 1866, as the oak budded and blossomed ; fed on oak blossoms and leaves ; full-grown June 2nd, 1866 ; imago end of July, 1866. 2 *a*, May 20th, 1871 ; imago July 25th, 1871. See pp. 127—129.

EUCLIDIA MI.

3, 3 *a*, larvæ in various stages of growth ; 3 *b*, larva coiled up ; 3, swept from marram, fed up on ribbon-grass, September 4th, 1872 ; imago June 13th, 1873 ; 3 *a*, 3 *b*, found on thorny rest-harrow ; ate grass ; figured August 25th, 1861 ; imago May 20th, 1862. See pp. 130—132.

EUCLIDIA GLYPHICA.

4, larva after last moult ; on sainfoin, clover, &c., August 30th, 1862. See pp. 132—134.

PHYTOMETRA ÆNEA.

5, 5 *a*, larvæ after last moult ; 5 *b*, cocoon ; 5 *c*, pupa ; on *Polygala vulgaris* (milkwort), September 1st, 1865, and July 29th, 1873. See pp. 134—137.

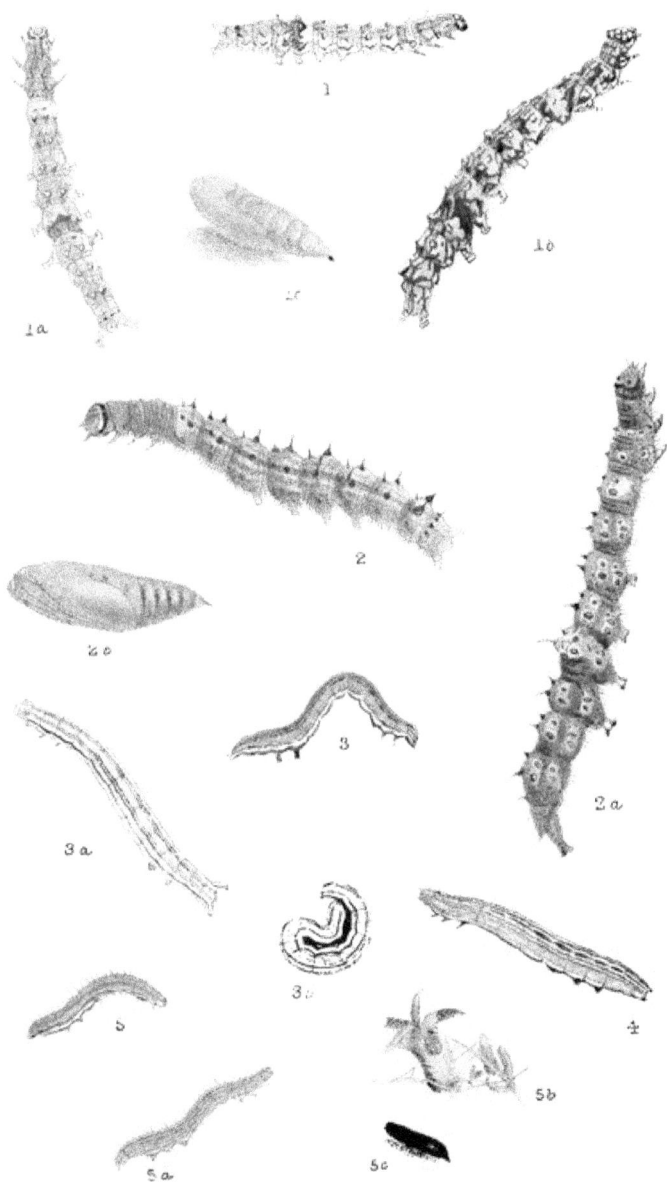

Plate CV.

F.W.Frohawk lith.

West, Newman imp.

W.BUCKLER del.

www.ingramcontent.com/pod-product-compliance
Lightning Source LLC
Chambersburg PA
CBHW021659210326

41599CB00013B/1468